Heat and Cold Storage 1

SCIENCES

Energy, Field Directors – Alain Dollet and Pascal Brault

Heat and Cold Storage, Subject Head – Philippe Marty

Heat and Cold Storage 1

Sensible and Latent Storage

Coordinated by
Pierre Odru
Elena Palomo del Barrio

ISTE

WILEY

First published 2024 in Great Britain and the United States by ISTE Ltd and John Wiley & Sons, Inc.

ISTE Ltd
27-37 St George's Road
London SW19 4EU
UK

www.iste.co.uk

John Wiley & Sons, Inc.
111 River Street
Hoboken, NJ 07030
USA

www.wiley.com

Library of Congress Control Number: 2023948927

British Library Cataloguing-in-Publication Data
A CIP record for this book is available from the British Library
ISBN 978-1-78945-133-7

ERC code:
PE8 Products and Processes Engineering
PE8_6 Energy processes engineering

Contents

Foreword

Philippe MARTY
LEGI, CNRS, Grenoble INP, Université Grenoble Alpes, Grenoble, France

There has always been an interest in storing heat: prehistoric people heating a stone using fire and then plunging it into a hollow piece of wood filled with water in order to cook their food can be thought of as conducting a first heat storage experiment. As for the Persians, they stored ice in the desert in dome-like buildings with thick clay walls.

Nowadays, heat storage is an important topic of debate on the development of renewable energies. It is generally accepted that massive deployment of these energies should be accompanied by significant storage facilities in order to cope with the irregularity of wind or insolation. Many industrial processes release waste heat. According to a study conducted by ADEME (French Agency for Ecological Transition), nearly 100 TWh of waste heat are released each year, which amounts to approximately one-third of the needs in this sector. In buildings, the currently developed heat storage methods rely essentially on storing water in balloons.

As for the national power grid, the massive solution that is most commonly implemented involves getting water at altitude and then passing it through turbines. Pumped hydroelectric energy storage (PHES) therefore stores electric energy in a gravitational form. This book will not discuss this technology. The same is true for electrochemical storage in batteries or hydrogen vector-based techniques that are the object of other volumes in the SCIENCES series of books published by ISTE.

Heat and Cold Storage 1,
coordinated by Pierre ODRU, Elena PALOMO DEL BARRIO.

The two volumes presented here have been written by renowned European experts in the field of heat storage and associated materials. They are exclusively dedicated to heat and cold storage and exhaustively discuss all the possible physical and technical solutions. Sensible heat storage methods, in which the temperature of a body is modified to store or release energy, will be presented. The technique involving phase change will also be presented in detail. Thermochemical or sorption heat storage, which is highly promising due to its larger density, will be dealt with in a separate volume.

Volume 1, *Heat and Cold Storage 1: Sensible and Latent Heat Storage*, focuses exclusively on these two types of energy transfers. The state of the art on the standard applications of storage for the building industry is presented, and long-term storage is given particular attention. Indeed, in a country with no steady insolation, the energy autonomy of a building depends on long-term storage possibility, which is also known as inter-seasonal storage. Underground storage offers in this respect possibilities that will be discussed.

The storage of electricity in the form of heat is studied in the presentation of new high-temperature processes. This promising technology is considered for very high power installations and can be seen as an alternative to PHES or to the use of hydrogen vector as a solution for the temporary storage of electricity.

Phase change materials (PCMs) are an attractive alternative when the temperature differences between the storage material and the external environment are too small. It is important to recall that the principle is using the phase change, most often solid–liquid, to accumulate or release the heat transferred in latent form. Due to the high value of the enthalpy of phase change of certain materials, compared to their heat capacity, storage density can be increased. A detailed presentation of the available materials is proposed. The wide range of applications of PCM for the storage of both heat and cold is rapidly extending. For example, they can be found in ice slurries in the form of encapsulated materials. The stakes in the micro- or nano-encapsulation process, and also in the impregnation of porous media, are huge, as they involve difficult techniques that are presented in detail in Volume 1. Besides a high mechanical and chemical stability, the thermal conductivity of the synthesized materials must be high enough to allow for reasonable energy transfer times. This requirement leads to the incorporation of highly conductive materials in the PCM in order to enhance their performances. Chapter 8 is dedicated to the applications of these materials in the industry and construction fields, including the particular case of storage integrated in solar power plants.

Cold storage is essential in a considerable number of applications, particularly in the food industry. Among the high-performance solutions, storage on hydrates is a future possibility that will be presented in Chapter 7.

The particularly promising case of concentrated solar power plants is also discussed in Chapter 8. Though current power plants essentially use molten salts, therefore relying on the principle of sensible storage, future developments of latent and thermochemical systems are presented.

Volume 2 is entitled *Heat and Cold Storage 2: Thermochemical Storage*. Similar to the electrochemical storage being able to sustainably store electrical power, thermochemical or sorption storage is able to store heat or cold over an a priori infinite duration. Unlike sensible or latent storage, this characteristic explains why energy engineers are very much in favor of this technique, which moreover presents a very high energy density that may reach several hundred kWh/m^3. This storage method in the form of chemical potential of the energy includes two distinct classes depending on whether the phenomenon in question is physical (van der Walls forces) or chemical (covalent bonds).

This volume introduces first the principles of sorption in a highly didactic manner and defines the criteria for the selection of materials to be implemented. This is followed by a detailed description of how an absorption cycle operates, which will indicate the future stakes of this method. The challenges that the researchers and manufacturers in the field are currently faced with relate to developing new designs of exchanger reactors, devising new cycles, and implementing new materials.

Particular attention is given to adsorption processes: after a review of the principles of this phenomenon and a thorough description of solid–gas adsorption cycles and systems, a large number of examples of prototype installations are presented.

These two volumes are addressed to several communities: researchers will find here a summary of the progress made by various universities in the field of storage, while academics will find a didactic presentation of the processes implemented. Engineers in research centers will find out the current technological stakes. There is room for improvement in many sectors: materials and associated manufacturing and characterization techniques, equipment, heat exchangers, new cycles, etc.

1

Sensible Heat Storage: Overview

Régis OLIVÈS
PROMES, CNRS, UPVD, Perpignan, France

1.1. Introduction

Heat and cold are two very important fields of the energy economy. They are used for the air-conditioning of living spaces, but they are also essential in the industry in the transformation and generation of materials and products. Nowadays, they are obtained from fossil fuels, either directly or using electricity, which is also often generated by power plants. But this combustion generates CO_2 emissions, contributing to the global warming, and these fuels are not sustainable, since they are not renewable, posing serious risks to the future development of humanity.

At the scale of humanity, sun is an inexhaustible source of heat. However, solar energy lacks concentration, it is intermittent and even lacking for a part of the year and its availability is often shifted with respect to the needs, particularly depending on the season. This explains the potential interest in the efficient storage of heat and also cold.

Two very different types of heat storage can be identified: low-temperature storage, the most widespread and mainly used for domestic air-conditioning and water heating, and high-temperature storage, more complex, essentially used to smooth power generation subjected to the intermittence of renewable energies. This work describes the processes of sensible heat storage, by far the most developed, but

Heat and Cold Storage 1,
coordinated by Pierre ODRU, Elena PALOMO DEL BARRIO.

does not cover the processes using phase change materials, which will be described in the following chapters, or the sorption processes, which will be dealt with in Volume 2.

1.2. General principles

Sensible heat is the heat stored in a material, except for phase changes (solid–liquid–gas transitions). Sensible heat storage in a material consists of its enthalpy increase during the storage phase, heat being released during the discharge phase. The material can be a liquid or a solid. The perceptible effect is temperature change. Stored energy is written as follows:

$$Q_{12} = m \int_{T_1}^{T_2} c_p\,(T) dT$$

Here, Q_{12} (Joule) is the sensible heat, $T_1$and T_2 (K) are the initial and final temperatures, c_p (J/kg·K) is the specific heat capacity of the body and it may be temperature dependent. If the specific heat capacity c_p remains constant, the relation (with ρ density, V volume) is written as follows:

$$Q_{12} = mc_p(T_2 - T_1) = \rho V c_p(T_2 - T_1)$$

It can be noted that for a given specific heat capacity and for a given amount of stored heat, the storage volume V is inversely proportional to the density of the material. These relations can be easily transposed to sensible cold storage. Sensible heat storage is by far the easiest to implement and also the oldest. If the material remains stable during temperature changes, the system is indefinitely reversible.

1.3. Storage configurations

There are various storage configurations that use liquid or solid materials for storage, in association or not with a gaseous or liquid heat-transfer fluid, carrying or not solid particles. The dimensioning depends on the type of configuration, operating temperatures, heat transfer modes involved, flow rates of the heat transfer fluid, thermophysical properties of the materials used, geometry of the storage tank and objectives in terms of energy and power. Depending on the storage technology, phenomena such as stratification may be predominant in the operation mode, dimensioning, management and efficiency of the storage.

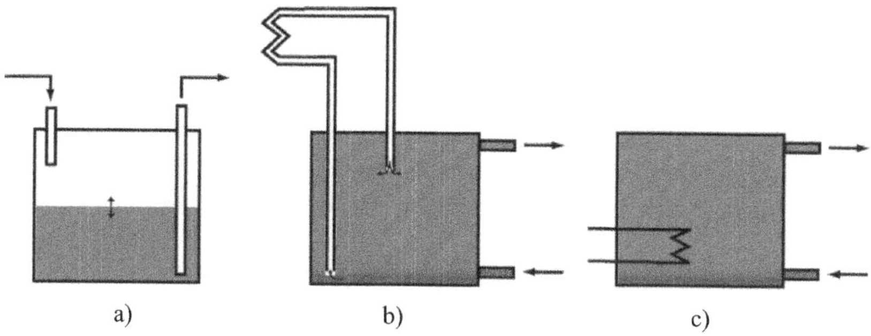

Figure 1.1. *Various sensible heat storage configurations. For a color version of this figure, see www.iste.co.uk/odru/heat1.zip*

Figure 1.1 shows three types of liquid storage configuration. In configuration (a), the storage involves filling the tank with heat-transfer fluid, which is also used as storage material. The release therefore involves emptying the tank and using the heat contained in the liquid. The recovered energy is therefore directly linked to the quantity of liquid and the available power is given by its circulating flow. After passage through an exchanger that may use the heat supplied by the storage, the liquid may then be stored at lower temperature in a so-called "cold" tank. Therefore, storage and release consist of the fluid passage from one tank to another. This configuration is currently used in concentrated solar power plants. The storage material is then quite often a mixture of molten salts, such as the binary mixture $NaNO_3/KNO_3$. In configuration (b), the storage consists of extracting the liquid from the bottom of the tank in order to heat it and then injecting it back at the top. The liquid is used both as heat-transfer fluid and storage material. The same is applicable to configuration (c) where, in contrast, a system placed inside the tank provides the heat to be stored. This can be done in various ways: flow of heat-transfer fluid, electric resistance, etc.

Configurations (b) and (c) are typically present in water heaters using electric power, solar power, etc. Specific studies have been conducted on these, particularly on stratification. For high temperature applications (>120°C), water is no longer the preferred fluid. Oils and molten salts are rather used. Given its high cost, a lower volume of liquid is used by associating it with inexpensive solid materials. At very high temperatures, air may prove to be the only viable technical solution. Considering the average air storage capacity, choosing solid materials for storage then becomes inevitable. Hence, the storage tank is filled with a solid material that takes up a part of the volume. The remaining volume (porosity) allows the heat-transfer fluid to supply with or extract the heat essentially stored in the solid material. This material can have a bulk granular form or an organized form. Granular materials of centimetric size can be natural rocks or ceramic materials (alumina, bauxite, recycled ceramic materials from industrial or municipal solid waste, etc.). Organized materials can be self-supported structured elements such as bricks or a material such as concrete, in which a set of tubes are inserted.

1.4. Modeling of thermocline storage

Operating this storage referred to as *dual media* involves the control of three zones: two high and low temperature zones separated by a high gradient zone, the thermocline.

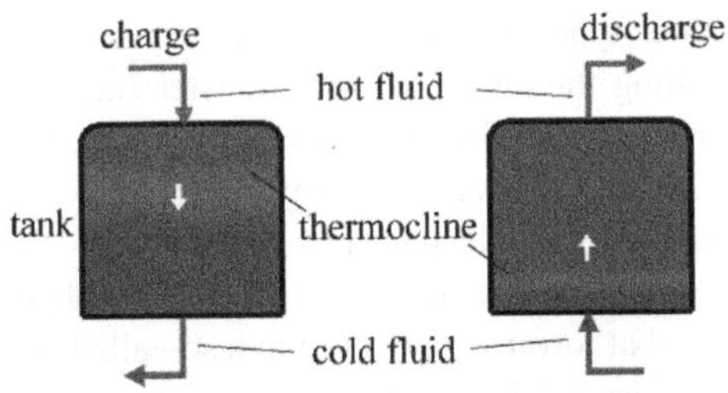

Figure 1.2. *Dual media heat storage. For a color version of this figure, see www.iste.co.uk/odru/heat1.zip*

The thermal modeling of storage on solid material relies on the coupling of convection heat transfer around the material and conduction transfer inside. There are two possible cases:

– The temperature of the medium can be considered homogeneous, and in this case the lumped heating or cooling method known as the *lumped capacitance method* can be used.

– There is a temperature distribution in the material, which requires complete or approximate calculation.

The first approach, very useful in storage, uses the lumped capacitance method. This highlights the influence of the material and of its thermophysical properties, and also of its geometry. Figure 1.3 shows a schematic representation of the lumped heating of a solid immersed in a fluid.

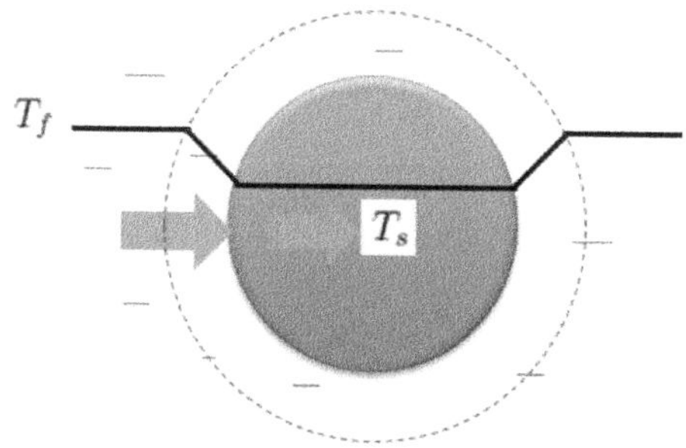

Figure 1.3. *Lumped heating of a solid by a fluid*

The energy supplied by the fluid is transferred by convection and leads to an increase in the internal energy of the solid:

$$h.S.(T_f - T_s).\mathrm{d}t = \rho.V.C_p.\mathrm{d}T_s \quad [1.1]$$

When the solid is at an initial temperature $T_{s,i}$ and it is subjected to a step in temperature at the initial moment, the result of integration is:

$$Ts = T_f + (T_{s,i} - T_f).\exp\left[-\left(\frac{h.S}{\rho.V.C_p}\right)t\right] \quad [1.2]$$

Let us consider $\theta = T_s - T_f$, $\theta_i = T_{s,i} - T_f$, which leads to the following expression:

$$\frac{\theta}{\theta_i} = \frac{T_s - T_f}{T_{s,i} - T_f} = \exp\left[-\left(\frac{h.S}{\rho.V.C_p}\right)t\right] \quad [1.3]$$

which yields:

$$\frac{\theta}{\theta_i} = \exp(-Bi.Fo) \quad [1.4]$$

The dimensionless form of this equation features Biot and Fourier numbers:

$$Bi = \frac{h.L_c}{\lambda} \quad [1.5]$$

and:

$$Fo = \frac{\alpha t}{L_c^2} \quad [1.6]$$

The characteristic length is the ratio between the volume of the required material and the surface through which transfer takes place:

$$L_c = \frac{V}{S} \quad [1.7]$$

This yields the following:

– For a plate of thickness e:

$$L_c = \frac{e}{2} \quad [1.8]$$

– For a cylinder of radius R:

$$L_c = \frac{R}{2} \quad [1.9]$$

– For a sphere of radius R:

$$L_c = \frac{R}{3} \quad [1.10]$$

Therefore, the stored energy is:

$$\frac{Q_{sto}}{Q_{max}} = 1 - \frac{\theta}{\theta_i} = 1 - \exp(-Bi.Fo) \quad [1.11]$$

It should be noted that this relation is also an expression of the storage efficiency.

These relations are valid for Biot numbers $Bi < 0.1$, therefore for thermally thin materials. Nevertheless, Xu et al. (2012) propose an extension of the lumped capacitance method to Biot numbers $Bi > 0.1$. An effective Biot number is introduced. In summary, for a flat plate, this effective Biot number is written as follows:

$$Bi_{eff} = \frac{Bi}{1+\frac{3}{5}.Bi} \quad [1.12]$$

For a cylinder, the correction is given by the following relation:

$$Bi_{eff} = \frac{Bi}{1+\frac{1}{2}.Bi} \quad [1.13]$$

Finally, for a sphere, we have:

$$Bi_{eff} = \frac{Bi}{1+\frac{1}{3}.Bi} \quad [1.14]$$

Generally speaking, the expression of the stored energy and therefore of the efficiency is given as follows:

$$\eta = \frac{Q_{sto}}{Q_{max}} = 1 - \exp(-Bi_{eff}.Fo) \quad [1.15]$$

This extension is interesting and quite precise in many cases. In the end, this situation corresponds to a storage that is limited by the convective transfer properties and therefore by the convective coefficient and the transfer surface. This explains the interest in granular medium as storage material. When the transfer in the solid is limiting ($Bi >> 0.1$), the equation of heat in the solid must be solved:

$$\frac{\partial^2 T_S}{\partial x^2} = \frac{1}{\alpha_S}\frac{\partial T_S}{\partial t} \quad [1.16]$$

for $x = L_c$:

$$\lambda_S \frac{\partial T_S}{\partial x} = h(T_f - T_s) \quad [1.17]$$

and for $x = 0$:

$$\frac{\partial T_S}{\partial x} = 0 \quad [1.18]$$

Then the general solution is:

$$\frac{T_S - T_f}{T_{S,i} - T_f} = \sum_{n=1,3,5...}^{\infty} A_n \exp\left[-\left(\frac{n.\gamma_n}{L}\right)^2 .\alpha.t\right] \cos\left(\frac{n.\gamma_n.x}{L}\right) \quad [1.19]$$

where the values of A_n and γ_n are tabulated (Bergman et al. 2011). For a Fourier number $Fo > 0.2$, the following approximations can be used:

– For a flat plate:

$$\frac{T_S - T_f}{T_{S,i} - T_f} = A_1 \exp(-\gamma_1^2.Fo)\cos\left(\frac{\gamma_1.x}{e}\right) \quad [1.20]$$

– For an infinite cylinder:

$$\frac{T_S - T_f}{T_{S,i} - T_f} = A_1 \exp(-\gamma_1^2.Fo)\mathrm{J}_0\left(\frac{\gamma_1.r}{R}\right) \quad [1.21]$$

– For a sphere:

$$\frac{T_s - T_f}{T_{s,i} - T_f} = A_1 \exp(-\gamma_1^2.Fo)\frac{\sin\left(\frac{\gamma_1.r}{R}\right)}{\frac{\gamma_1.r}{R}} \quad [1.22]$$

For Fourier numbers $Fo < 0.2$, Heissler charts may prove useful (Bergman et al. 2011).

Let us now integrate to the storage the transfer between the material and the flowing fluid. It is typically the case of thermocline storage that is used in industrial waste heat recovery, in concentrated solar power plants, or still in the glass industry, being then called regenerators. Figure 1.4 shows the charge of such a storage.

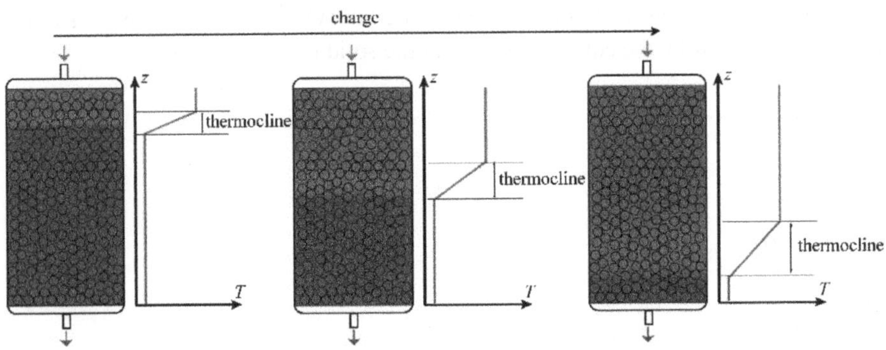

Figure 1.4. *Charge of a packed granular bed thermocline storage system. For a color version of this figure, see www.iste.co.uk/odru/heat1.zip*

Willmott (2001) established the equations that model this storage using two coupled equations: it is a two-temperature model. Two large types of models can be distinguished: a 1D-1D model and a 1D-2D model. Esence et al. (2029) present all the necessary elements on the numerical differences of models.

An illustration of the 1D-1D model is Schumann's (1926) model. This model, used once again by Hughes et al. (1976), was established for a particle bed. The effective conductivity of the bed is assumed to be infinite in the radial direction; there is no temperature gradient in the particles; the fluid flows in a single direction; the conductivity of the particle in the axial direction is zero; the fluid has no axial thermal dispersion; the properties do not depend on temperature; no heat loss in the environment is considered. The radiative transfer between fluid and solid and between the particles in the bed is negligible.

Then the model is written as follows:

– In the fluid phase:

$$\varepsilon C_{pf}\rho_f \frac{\partial T_f}{\partial t} = -u_f \varepsilon C_{pf}\rho_f \frac{\partial T_f}{\partial x} + h_c A(T_s - T_f) \qquad [1.23]$$

where A is the surface of a particle and ε is the porosity.

The interstitial velocity v_f can be written using the mass flow rate. Then we have:

$$u_f = \frac{v_f}{\varepsilon} \qquad [1.24]$$

$$v_f = \frac{\dot{m}_f}{\rho_f A_{cs}} \qquad [1.25]$$

where A_{cs} is the fluid passage section area.

– In the solid phase:

$$(1-\varepsilon) C_{ps}\rho_s \frac{\partial T_s}{\partial t} = h_c A(T_f - T_s) \qquad [1.26]$$

The system of equations is accompanied by the initial conditions, which are often a uniform temperature and the boundary conditions that can be a jump in the temperature of the ingoing fluid.

Let us introduce a volume exchange coefficient $h_v = h_c.A$. Then the number of transfer units (NTU) associated with the particle bed can be defined as follows:

$$NTU = \frac{h_v A_{cs} L}{\dot{m}_f C_{pf}} \qquad [1.27]$$

For a low heat capacity fluid such as air, the term $\rho_f C_{pf} \partial T_f / \partial t$ can be ignored. Then the equation related to the fluid becomes:

$$\frac{\partial T_f}{\partial x} = \frac{h_v}{v_f \rho_f C_{pf}}(T_s - T_f) = \frac{h_v A_{cs} L}{\dot{m}_f C_{pf} L}(T_s - T_f) \qquad [1.28]$$

hence:

$$\frac{\partial T_f}{\partial x} = \frac{NTU}{L}(T_s - T_f) \qquad [1.29]$$

Duffie and Beckman (2006) introduced a time constant:

$$\tau = \frac{\rho_s C_{ps}(1-\varepsilon)A_{cs}L}{\dot{m}_f C_{pf}} = \frac{m_s C_{ps}}{\dot{m}_f C_{pf}} \quad [1.30]$$

This time constant is characteristic for the ratio of heat stored by the solid and transferred by convection by the fluid. Then we have:

$$\frac{\partial T_s}{\partial t} = \frac{NTU}{\tau}(T_f - T_s) \quad [1.31]$$

If there is a temperature gradient through the particles ($Bi_D > 0{,}1$), Duffie and Beckman (2006) propose a modified NTU:

$$NTU_{eff} = \frac{NTU}{1+Bi_D/5} \quad [1.32]$$

with:

$$Bi_D = \frac{h_c D}{2\lambda_s} \quad [1.33]$$

The Schumann model can be completed by a term describing axial diffusion. Hence:

$$\varepsilon C_{pf}\rho_f \frac{\partial T_f}{\partial t} = \lambda_{eff,f}\frac{\partial^2 T_f}{\partial x^2} - u_f \varepsilon C_{pf}\rho_f \frac{\partial T_f}{\partial x} + h_c A(T_s - T_f) \quad [1.34]$$

and:

$$(1-\varepsilon)C_{ps}\rho_s \frac{\partial T_s}{\partial t} = \lambda_{eff,s}\frac{\partial^2 T_s}{\partial x^2} + h_c A(T_f - T_s) \quad [1.35]$$

where $\lambda_{eff,f}$ and $\lambda_{eff,s}$ are the effective conductivities of the fluid and solid, respectively.

If, from now on, radial dispersion is considered in the solid, the model becomes 1D-2D (1D fluid, 2D solid). This model was implemented in the works conducted particularly at PROMES-CNRS Laboratory and at CEA-Liten (Couturier et al. 2014; Hoffmann 2015; Kéré et al. 2015; Esence et al. 2017; Fasquelle 2017; Lopez-Ferber 2018; Touzo 2021) to describe heat storages at relatively high temperatures (from 200 to 850°C).

Finally, some remarks can be made on the performances of sensible heat storages. The group of dimensionless numbers that are naturally involved in this analysis is $Bi.Fo$. It corresponds to the ratio between transferred power and stored power. $Bi.Fo$ can also be expressed as the product of the number of transfer units NTU and the ratio between the energy transferred by convection by the fluid) and the energy stored by the solid. Then we introduce the Stanton number St that compares the heat transferred by convection to the heat carried by the flowing fluid. NTU is then proportional to the Stanton number via the ratio between the transfer surface and the flow section:

$$NTU = St\frac{S_{ech}}{S_{eclt}} \tag{1.36}$$

The group $Bi.Fo/St$ can be formed as follows:

$$\frac{Bi.Fo}{St} = \frac{h.V_S}{\lambda_S.S_{ech}}.\frac{\lambda_S.t.S_{ech}^2}{\rho_S.C_{pS}.V_S^2}.\frac{\rho_f.C_{pf}.u_f}{h} = \frac{\rho_f.C_{pf}.u_f.t}{\rho_S.C_{pS}.V_S}.S_{ech} = \frac{\dot{m}_f.C_{pf}.t}{m_S.C_{pS}}.\frac{S_{ech}}{S_{eclt}} \tag{1.37}$$

This introduces the ratio between the energy transferred by convection by the fluid and the stored energy. In other terms, this group is characteristic for the time required for the fluid to extract or supply the storage energy taking into account the available transfer surface/flow section ratio. This highlights four dimensionless numbers, Bi, Fo, St and S_{ech}/S_{eclt}, that can be used to analyze and qualify the sensible heat storage.

1.5. References

Bergman, T.L., Lavine, A.S., Incropera, F.P., Dewitt, D.P. (2011). *Fundamentals of Heat and Mass Transfer*. John Wiley & Sons, Hoboken.

Couturier, R., Bruch, A., Fourmigué, J.F. (2014). Experimental and numerical investigation of a pilot-scale thermal oil packed bed thermal storage system for CSP power plant. *Solar Energy*, 105, 116–125.

Duffie, J.A. and Beckman, W.A. (2006). *Solar Energy Thermal Processes*. John Wiley & Sons, New York.

Esence, T. (2017). Étude et modélisation des systèmes de stockage thermique de type régénératif solide/fluide. PhD Thesis, Université Grenoble Alpes, Grenoble.

Esence, T., Desrues, T., Fourmigué, J.-F., Cwicklinski, G., Bruch, A., Stutz, B. (2019). Experimental study and numerical modelling of high temperature gas/solid packed-bed heat storage systems. *Energy*, 180, 61–78.

Fasquelle, T. (2017). Modélisation et caractérisation expérimentale d'une boucle solaire cylindro-parabolique intégrant un stockage de type thermocline. PhD Thesis, Université de Perpignan Via Domitia, Perpignan.

Hoffmann, J-F. (2015). Stockage thermique pour centrale solaire thermodynamique à concentration mettant en œuvre des matériaux naturels ou recyclés. PhD Thesis, Université de Perpignan Via Domitia, Perpignan.

Hughes, P.J., Klein, S.A., Close, D.J. (1976). Packed bed thermal storage models for solar air heating and cooling systems. *J. Heat Transfer*, 98, 336–338.

Kéré, A., Goetz, V., Py, X., Olives, R., Sadiki, N. (2015). Modeling and integration of a heat storage tank in a compressed air electricity storage process. *Energy Conversion Management*, 103, 499–510.

Lopez-Ferber, N. (2018). Validation expérimentale d'un système de stockage thermo-cline air/céramique à échelle pilote - développement d'un matériau de stockage. PhD Thesis, Université de Perpignan Via Domitia, Perpignan.

Schumann, T.E.W. (1929). Heat transfer: A liquid flowing through a porous prism. *Journal of the Franklin Institute*, 208, 405–416.

Touzo, A. (2021). Intégration d'un stockage thermocline à des procédés énergétiques. PhD Thesis, Université de Perpignan Via Domitia, Perpignan.

Willmott, A.J. (2001). *Dynamics of Regenerative Heat Transfer*. Taylor & Francis, London.

Xu, B., Li, P.-W., Chan, C.L. (2012). Extending the validity of lumped capacitance method for large Biot number in thermal storage application. *Solar Energy*, 86(6), 1709–1724.

2

Low-Temperature Sensible Heat Storage

Pierre ODRU
Previously IFPEN, Paris, France

Unlike fossil fuels that can be easily and indefinitely stored, heat is an energy difficult to store. Solar heat, which can be captured by solar thermal panels as hot water, is essentially available during summer, though it is most needed during winter. The same for cold available during winter. This explains the interest in reliable and large-scale storage solutions that may also allow the collection of significant amounts of heat wasted in industrial processes. Storage directly associated with buildings will be distinguished from underground storage.

2.1. Sensible heat storage associated with buildings

Sensible heat storage associated with buildings stores the heat collected by solar thermal panels over various time scales, from several days to transferring the heat collected during the hot season to the cold season (vice versa for cold). The required properties of the storage material are high specific and volumetric heat capacities, to reduce the necessary masses and volumes, and also good thermal conductivity to facilitate exchanges.

For the temperatures required in building (heating, cooling, domestic hot water), concrete or building materials can be used passively. But water is particularly adapted and can be directly circulated in order to provide domestic hot water or heating. It is moreover economical and neutral to the environment. It may however

Heat and Cold Storage 1,
coordinated by Pierre ODRU, Elena PALOMO DEL BARRIO.

facilitate corrosion and it has a limited use temperature domain, though well adapted to the building.

Material	Temperature (°C)	Density (kg/m³)	Specific heat capacity (J/kg·K)	Volumetric heat capacity (kJ/m³·K)	Thermal conductivity (W/m·K)
Water	0–100	1,000	4,190	4,190	0.63
Ice	<0	920	2,060	1,895	2.1 (0°C)
Concrete	0–150	2,250	1,130	2,542	0.9–1.3
Brick	–	1,700	850	1,400	0.7

Table 2.1. *Thermal characteristics of materials*

The disadvantage of sensible heat storage is its low-energy density, requiring large volumes. This is why standard tanks have limited capacity. Very large capacities require the use of external storage or geological structures, such as aquifers, rocks or soils under certain conditions.

In general, heat extraction must be conducted so that output temperature is maintained as constant as possible. For example, in the case of domestic hot water, if three quarters of a storage at 90°C are used, the user experience is not the same if temperature stays at 90°C or if it is 37°C if the already used hot water is replaced by water at 20°C with mixture. This generally requires temperature stratification, which is difficult to maintain, and increases system complexity.

2.1.1. *Short duration storage*

Solar heat collected by solar thermal panels is essentially available during the day, and in a temperate country it may be absent for several days in a row. This justifies short duration storage. The principle of these storage facilities is well known. They consist of metallic or fiberglass composite tanks coated with adapted insulation material, glass wool, polystyrene or polyurethane foam. The calculations allowing the dimensioning of these types of tanks are quite standard. The ageing of materials must obviously be known and adapted.

Typically, for a single-family home in Europe (two to five persons), a solar water heater dedicated exclusively to domestic hot water is composed of solar collectors with a total surface of around 4 m² and a storage tank of 200–300 L. The solar coverage rate of such an installation, the part of water heated by solar energy, ranges between 60 and 90%.

In temperate countries, solar water heaters cannot by themselves meet the needs and must be used in combination with a classic gas or electric water heater, especially if they combine the production of domestic hot water and the underfloor heating of the building during winter. Some manufacturers propose using an auxiliary wood-firing boiler.

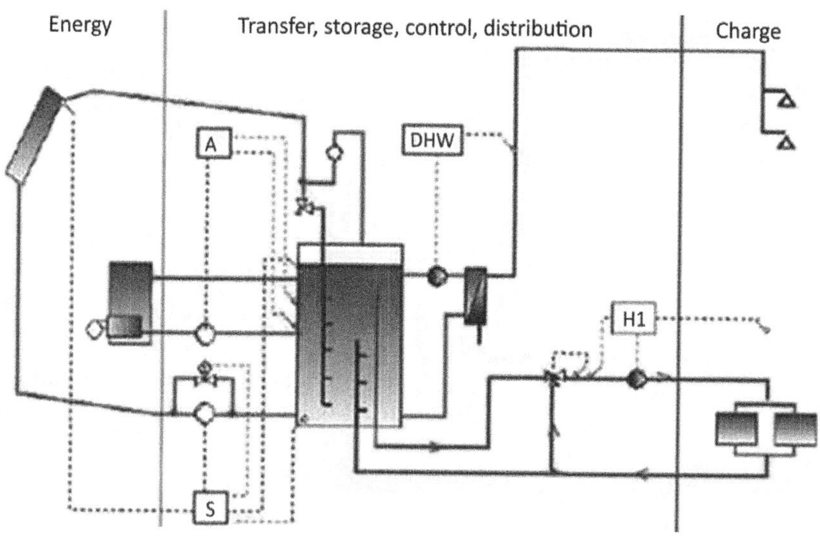

Figure 2.1. *Maxlean combined heating system. For a color version of this figure, see www.iste.co.uk/odru/heat1.zip*

In some cases, water can be used in an open loop, but the heat from the solar thermal panel is generally transmitted from an exchanger located, for stratification reasons, in the lower (cold) part of the storage tank (Hadorn 2005). The heat is then recovered by a second exchanger bathing in the upper (hot) part of the tank and distributed in the building. As part of the SHC (Solar Heating and Cooling) program of the IEA (International Energy Agency), a combined system design, known as Maxlean concept, was subjected to an advanced energy and economic optimization.

The schematic representation in Figure 2.1 shows the following:

– the energy supply zone: solar collectors and classic water heater;

– transfer, storage, control and distribution zone;

– charge, namely domestic hot water and home heating.

All the processes allowing the optimization and also the sensitivity studies are provided in IEA SHC Program (2008a).

2.1.2. *Long duration (seasonal) solar heat storage*

Seasonal solar heat storage obviously requires large volumes of water that are heated during summer to be used in winter. Large size tanks have the advantage of reduced surface to volume ratio, which limits thermal losses, but the duration being significant, heat insulation must be important. This storage should be obviously associated with buildings having high-quality insulation and sufficient solar collector surface allowing heat storage to be completed by the end of summer and to last the longest time possible during winter. An additional heating system is however generally needed.

The storage volume has an important place in the design of the building or house. But it can also be placed outside and underground in the case of a building or a group of buildings. It can be made from reinforced concrete with a sealing liner. The soil may play the role of additional insulator, but it will first absorb a part of the heat before the system reaches equilibrium.

Building such a tank obviously requires taking into account average insolation conditions, the efficiency of solar panels, the heat insulation of the buildings and the tank, pumping systems and consumption.

According to a Swiss study (Hadorn 2005), a 200 m^2 house with high-quality insulation would only require 15 m^2 of collectors to meet the domestic hot water and heating needs of 6,100 kWh/year. However, half of the needs occur during the three winter months when solar heat input is low. Under these conditions, the storage transfer needs would be 1,850 kWh. Only 10 m^2 of additional collectors would then be required, without counting heat losses.

For actual examples, please refer to IEA Preheat Programme (2007).

2.1.3. *The stratification problem*

A storage system with an exchanger consists of an actual storage volume, in which the cold water, heavier, is supplied by the bottom, and the hot water, lighter, is collected at the top. The exchanger bringing the heat from solar panels is at the bottom and heats the cold water that goes up due to the density difference in the tank and accumulates at the top part.

Water stratification, due to the variation of density with temperature, will therefore play a critical role in system efficiency, with an intermediate zone in which heat exchanges take place. The system is all the more efficient as this stratification can be maintained for a longer time and it gives rise to a more significant temperature gradient.

Reaching proper stratification is a difficult task. For efficiency reasons, a high and thin storage tank is preferable. If water flows through the storage tank, hot water can be injected at the most favorable place in order to limit water displacement and thus increase the efficiency of stratification. This can be achieved using a vertical pipe with opening valves at various levels, electronically controlled to open at the desired level (see the Maxlean system). A flexible pipe with a density allowing it to float at the expected level can also be used. There are various methods for calculating stratification in a tank (see for example Haghighat (2013, p. 70) and Chapter 1).

2.2. Underground thermal energy storage

Another method for seasonal heat (or cold) storage for residential or tertiary applications involves the use of large underground storage capacities, which offer free on-site materials and favorable thermal characteristics.

Material	Temperature (°C)	Density (kg/m³)	Specific heat capacity (J/kg·K)	Volumetric heat capacity (kJ/m³·K)	Thermal conductivity (W/m·K)
Water	0–100	1,000	4,190	4,190	0.63
Compact rock	–	2,600	800	2,080	2.5
Dry soil	–	1,300	800	1,050	1.28
Wet soil	<100	1,700	2,000	3,400	0.52

Table 2.2. *Thermal characteristics of rocks and soils*

Beyond a certain depth, the basement has a constant temperature, equal to the average of the annual temperature on the site, and it also has significant thermal inertia, coupled with very significant volumes of available materials. Heat and cold can be alternatively stored in it, depending on the season. However, the fact that soils and rocks may contain water generally limits their range of use at temperatures below 100°C.

There are essentially aquifer thermal energy storage (ATES) and borehole thermal energy storage (BTES), which store by conduction in the solid medium, and finally subsurface energy storage, using foundations. The heat transfer fluid used is in general glycol water to eliminate freezing risks.

A geological heat storage system requires three conditions to be met at the same location:

– energy resource: solar hot water collected in summer, waste heat from industrial installations, cool water stored in summer, etc.;

– energy demand shifted in time: need for heating buildings during winter, need for cooling during summer;

– geological and hydrogeological context favorable to underground storage of heat or cold.

Investment is significant and must be economically justified in the long term.

Moreover, the system should have auxiliary heating or cooling systems (heat pumps, classic heating or cooling, etc.). In addition to these, there is an obligation for sustainable management of the environmental impact, including the preservation of the integrity of underground drinking water.

Projects will depend also on the proper evaluation of the heating and cooling demand of the buildings, and on the proper dimensioning of the surface installations: coefficient of performance (COP) of the heat pumps, thermal insulation, proper dimensioning of pipelines, energy efficiency of the pumps, heat exchangers, efficiency of the regulation systems, etc.

Storage is all the more efficient as the losses by thermal diffusion in the underground remain relatively limited, and therefore as it is larger. Thermal equilibrium in the underground is only reached after several years, inducing more significant losses during the first period of operation. There may be a great impact on the return on investment.

2.2.1. *Principle of the aquifer thermal energy storage*

Aquifer thermal energy storage is possible only if the aqueous phase is not flowing, which requires specific geological conditions and limits the potential. It can compete with other uses.

This system requires at least two distant wells to prevent cold and hot parts from mixing. Exchanges are also possible between distinct aquifers (Reichstag in Berlin), a cold one near the surface and a hot one at depth.

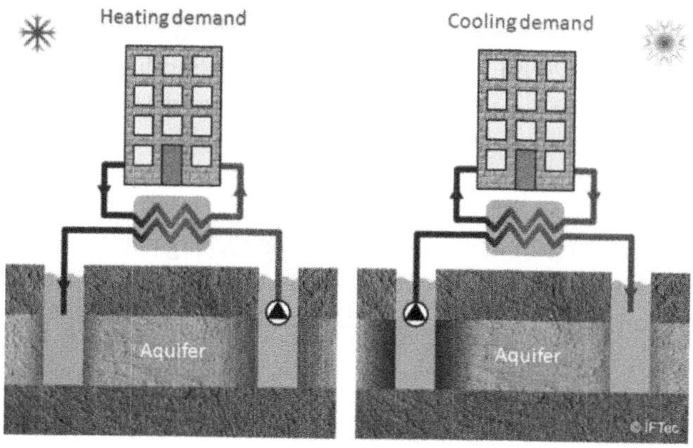

Cold well Hot well Cold well Hot well

a) b)

Figure 2.2. *ATES closed loop (IEA-DHC 2018). (a) Winter mode; (b) Summer mode. For a color version of this figure, see www.iste.co.uk/odru/heat1.zip*

In the simplest case, cold water is pumped in summer and is heated through an exchanger by a thermal solar source or by cogeneration and is sent to the aquifer. It then heats the building in winter, and once cooled it is sent back to the aquifer. In many cases, cold water pumped during summer can also be used for cooling the building, before receiving the complementary thermal input.

A heat pump may complete the installation, in addition to the classic heating or cooling means.

Reversible heat exchanges are not only convective through water, but also mobilize a diffusive part in the neighboring solid layers. Reaching thermal equilibrium may take several years, which entails the initial loss of significant amounts of heat. Available power depends on the pumped flow rate and on the quantity of stored hot water. Once the equilibrium is reached, the recovered to injected energy ratio may exceed 75%; thermal storage then has a dominant part in the heating balance, or of electric power consumption when cooling is involved.

Basements in the Netherlands are very favorable to the implementation of ATES systems, which allowed a certain number of local enterprises to gain highly competitive know-how. The technology is considered mature. There is fast return on investment in favorable cases.

A well-documented ATES example, from both technical and economic perspective, is that of cold storage conducted at Richard Stockton College in New Jersey (Packsoy et al. 2009). Wells and their installation have considerable costs. These may however be lowered after a learning period. A significant return on investment is expected after 20 years of operation, with a decrease in the carbon footprint of the institution.

A study was conducted in France on off-season recharge of a deep aquifer used in geothermal system with hot water from a household waste incineration plant. The objective was to increase the winter potential and avoid resource depletion (Reveillere et al. 2013).

The Parisian Basin has an interesting geothermal heat potential, which happily coincides with strong local demand. The most interesting and widely exploited geological formation is that of Dogger. Water temperature varies there between 55 and 85 °C for depths ranging between 1,500 and 2,000 m. A 200 m^3 per hour productivity can be reached for limestone levels showing fracturing and dissolution. The water contained in the aquifer is however highly mineralized. Geothermal exploitation of the resource requires deep dual geothermal drilling. Two wells are drilled at 45° from the surface and thus reach the formation in two points separated by a distance of about 1,700 m. Hot water is pumped by one of the wells, brought to the surface, where it exchanges its heat before being reinjected in a cooled state in the second well. The heat generated is then collected by an existing local heating network.

To the right of the injection point a cold bubble is formed, which grows larger during the years. When the cold bubble reaches the pumping well, the geothermal reserve is depleted. The geothermal resource is in fact not sustainable, as the amount of heat released by the earth is significantly smaller than the exploited rate. The lifetime of a dual geothermal drilling is estimated at approximately 30 years.

This explains the interest in finding heat resources that may recharge the formation during summer and allow the geothermal resource to become renewable. This is possible in Île-de-France by the exploitation of hot waters produced by the household waste incineration plants in the vicinity of geothermal wells. A plant injects annually 300,000 oil equivalent in urban heating in winter and 100 MW power is available in summer during the low demand period. Paris region has a heating network that supplies over 1 million inhabitants, a significant part of the energy supply being geothermal. Therefore, the idea is to inject hot water in the formation when out of the cold period in order to increase the amount of heat that can be recovered in winter.

In contrast with a classic geothermal exploitation, in summer, the cold well is used as a generator, its cold water being heated at the surface in contact with the hot water from the plant, then reinjected at 95°C in the hot well, thus increasing the amount of available hot water and also its temperature. In winter, the hot well starts producing, injecting its heat via an exchanger in the heat network, then injects cold water in the cold well. At the end of winter, the injected hot water is exhausted, while the amount of cold water injected increases, which means that both the heat stored during summer and the geothermal potential of the formation are used.

The tank is modeled by a single equivalent layer in a hosting rock. The characteristics of Dogger in Île-de-France are well documented. Calculation software allowing the simulation of various types of exploitation of the deposit (summer injection, winter production, pauses between the two) have been developed and allow the evaluation of production, the evolution of temperatures and the evolution of the size of the zones of the deposit impacted by the exploitation, particularly the increase in the size of the cold zone during the decades of exploitation (Reveillere et al. 2013). This can then be translated in the exploitation of the heat network.

2.2.2. *Borehole thermal energy storage*

A field of boreholes consists of wells drilled in a rocky massif, generally based on vertical concrete U tubes. The heat-transfer fluid goes in hot and comes out colder (or vice versa). In a reversible system, cold water intended to cool the building is pumped in summer and sent back heated by solar collectors. Conversely, in winter, hot water is pumped and sent back in the underground after having been cooled. A heat pump can be added. Heat is transferred by thermal diffusion in the rocky massif. The most important parameters for the dimensioning of the underground system are the thermal conductivity of the hosting ground and the efficiency of the field of boreholes in terms of heat transfer. Obviously, no water flow must be present.

Efficiency is generally all the higher as a compact hosting rock is available, with a high specific heat capacity, and also with a rather average thermal conductivity (to prevent heat from escaping to neighboring grounds). The thermal conductivity of the underground medium depends on the size of the rocky blocks, fractures, water saturation and the intrinsic properties of the materials composing the rock.

The field of boreholes may have various geometries, but it must remain compact and meet adequate dimension ratios, which partly depend on the geological characteristics, in order to have an efficient storage and avoid losses. The available power depends on the thermal conductivity of the medium and also on the surfaces in contact, therefore on the number of boreholes, their depth, their diameter, the materials injected in the borehole and their arrangement. The thermal equilibrium conditions vary during the operation cycles, as the exchanges heat or cool the massif. The return on investment of a BTES is generally longer than the one of ATES, but the compatible geological sites are more numerous and the associated risks are fewer.

The field of boreholes in Drake Landing Solar Community (Alberta, Canada) is an example of very well documented BTES installation, with heat supplied by solar panels aimed at covering over 90% of the heat needs of the community.

Fifty-two single-family homes with high energy efficiency standards are connected. The heat resource is provided by 800 collectors mounted at 45° on the garages and covering a 2,300 m^2 surface. The peak thermal power is 1.6 MW, and the heat-transfer fluid is glycol water. The domestic hot water circuit is separate. The seasonal storage (BTES) and the short-term storage (night) are controlled by two units.

The general efficiency of BTES has improved in time and has reached 54% in 2011. In 2012, the direct use of the heat supplied by solar panels covers 54% of the final consumption, the heat provided by BTES storage covers 43% and gas heating only 3%.

A new BTES design was proposed at average or great depth (Welsch et al. 2016) that may reach very high storage efficiency values. The boreholes are deeper, up to 1,000 m compared to the current BTES whose depth does not exceed 100 m. Their surface footprint would be less significant, which may play an important role in urban areas, and their depth may allow a decrease in heat losses and interactions with surface aquifers.

The projects would thus depend less on geology. But deep drilling technology is more complex and expensive, and economic benefit should be proven.

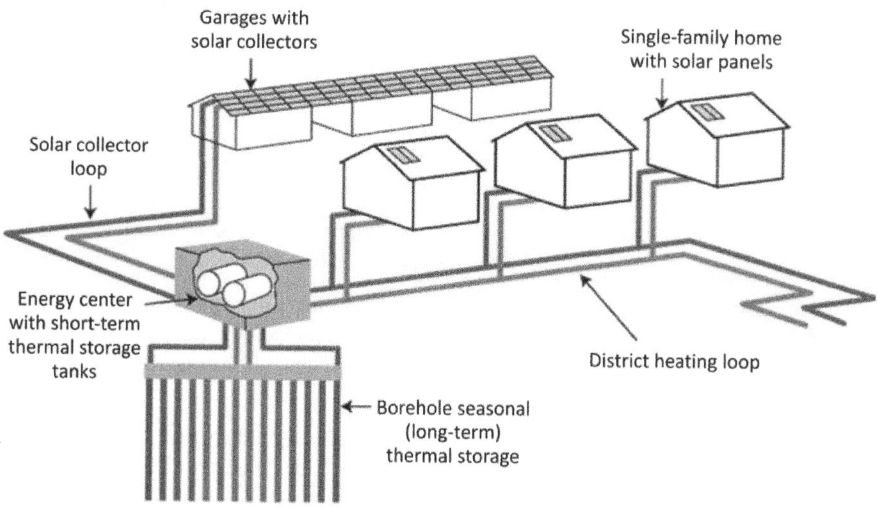

Figure 2.3. *Drake Landing Solar Community storage system: general layout*[1]*. For a color version of this figure, see www.iste.co.uk/odru/heat1.zip*

2.2.3. *Other types of underground storage*

Heat or cold storage in the foundations or geostructures becomes increasingly important, as it can be easily achieved with the support of expert design engineering (integration required at the design level). The structure of the foundation is equipped with pipes through which flows the water exchanging heat or cold, depending on the season, with the surrounding ground. Quite obviously, the storage function should by no means alter the structural function, allowing no freezing risk and requiring an annual return to initial conditions.

Heat or cold storage, potentially by accumulation of snow, can also be achieved by cavern thermal energy storage (CTES).

2.3. References

Drake Landing Solar Community (n.d.). Welcome to Drake Landing Solar Community [Online]. Available at: www.dlsc.ca/.

Hadorn, J.C. (ed.) (2005). Thermal energy storage for solar and low energy buildings. International Energy Agency.

1. Available at: www.dlsc.ca.

Haghighat, F. (2013). Applying energy storage in ultra-low energy buildings. Final report, International Energy Agency.

IEA Preheat Programme (2007). Solar storage in district heating networks. International Energy Agency.

IEA SHC Programme (2008a). 2008 Annual Report. International Energy Agency Solar Heating & Cooling Programme [Accessed 25 October 2023].

IEA SHC Programme (2008b). Solar simulation and optimization report – MaxLean concept. International Energy Agency Solar Heating & Cooling Programme.

IEA-DHC (2018). Integrated cost-effective large-scale thermal energy storage for smart district heating and cooling: Design aspects for large-scale aquifer and pit thermal energy storage for district heating and cooling [Online]. Available at: iea-dhc.org [Accessed 21 November 2023].

Lesueur, H. (2015). Les stockages géologiques de l'énergie thermique – Réalités et promesses. In *IDées ENR et Stockage de l'Energie Seminar*, Fondation Tuck.

Nordell, B. (2012). Underground thermal energy storage (UTES). In *Innostock 2012: The 12th International Conference on Energy Storage*, Lleida.

Packsoy, H., Snijders, A., Stiles, L. (2009). Aquifer thermal energy cold storage system at Richard Stockton College. In *Effstock 2009, Thermal Energy Storage for Efficiency and Sustainability: 11th International Conference on Thermal Energy Storage*, Stockholm.

Reveillere, A., Lesueur, H., Hamm, V., Cordier, E. (2013). Geothermal contribution to the energy mix of a heating network when using aquifer thermal energy storage: Modeling and application to the Paris basin. *Geothermics*, 47, 69–79.

Welsch, B., Ruhaak, W., Schulte, D., Baer, K., Sass, I. (2016). Characteristics of medium deep borehole thermal energy storage. *International Journal of Energy Research*, 40(13), 1855–1868.

3

High-Temperature Heat Storage for Electricity

Lionel NADAU[1], Philippe MUGUERRA[2] and Pierre ODRU[3]

[1] *ENGIE, Stains, France*

[2] *Saipem, Montigny-le-Bretonneux, France*

[3] *Previously IFPEN, Paris, France*

Solar heat is an inexhaustible source of renewable energy, but it is diffuse and intermittent, particularly between day and night, or winter and summer. In order to reach its full potential, large-scale storage should be required. But heat (or cold) can also be used as a storage medium or energy vector for the generation of electricity using intermittent wind and photovoltaic renewable energy. Unlike fossil fuels, which can be readily stored and require a thermodynamic process to provide usable energy, electricity is an energy that cannot be stored directly. The amount generated at each moment must be equal to the demand, with the risk of a frequency variation exposing the grid to difficulties, even to collapse. The demand, which is lower during the night, increases in the morning until reaching a plateau, and then decreases in late afternoon and increases in the evening, particularly during winter in France, where there are numerous single household electric heating systems (Figure 3.1). In the old scenario, fossil fuel (coal, gas) or nuclear power plants, and hydroelectric power plants along rivers, meet the basic needs. The peak and off-peak demand are then absorbed by a modulation of these power plants and supplements provided by gas turbines, or by hydroelectric energy collected in dams and released when needed.

Heat and Cold Storage 1,
coordinated by Pierre ODRU, Elena PALOMO DEL BARRIO.

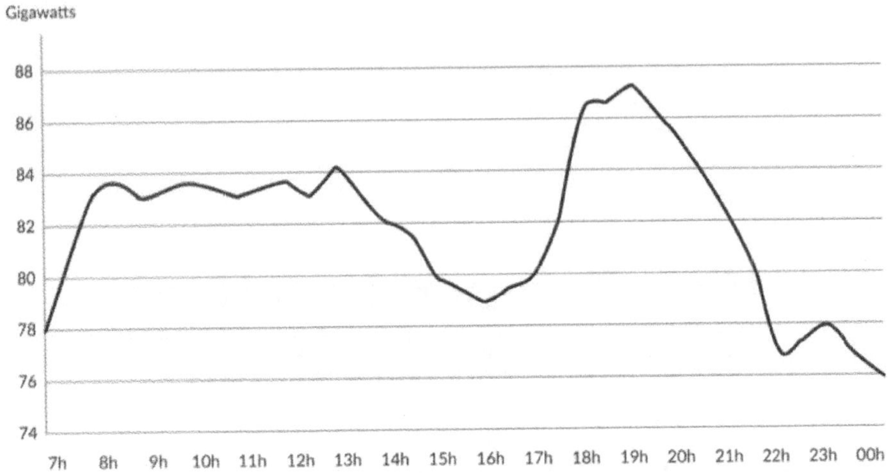

Figure 3.1. *Daily consumption of electricity in France (source: RTE)*

The complexity of this scenario increases when renewable energies are introduced. Unlike controllable thermal or nuclear power plants, wind-generated or photovoltaic electricity introduced in the grid, having priority in our countries is intermittent (but predictable) and not available on demand (e.g. there is no sun during the night, so it is therefore difficult to generate electricity with solar panels). While the forecast and management of highly interconnected networks allow their proper absorption when the injected quantities are low, it is not the case when the grids are small or when there is a significant penetration rate of renewable energies.

In Northern Europe, the highest potential renewable energy is wind power. According to the experience in France, the average yearly production (except offshore) is below a quarter of the installed power, with very significant variations, as indicated in Figure 3.2.

It is therefore obvious that if complementary supply is not available, significant surpluses or shortages may prove very harmful for both consumption and grid, with *black-out* risks that may be highly detrimental to economy and security. Among the many solutions, direct massive storage of electricity would be the best tool, if it was possible (Odru et al. 2013). Besides multiplying sources and various complementarities, other solutions are interregional transportation, exchanges between countries, shifting of demand when possible, intelligent networks, in addition to hydroelectricity and fossil energies as a last alternative.

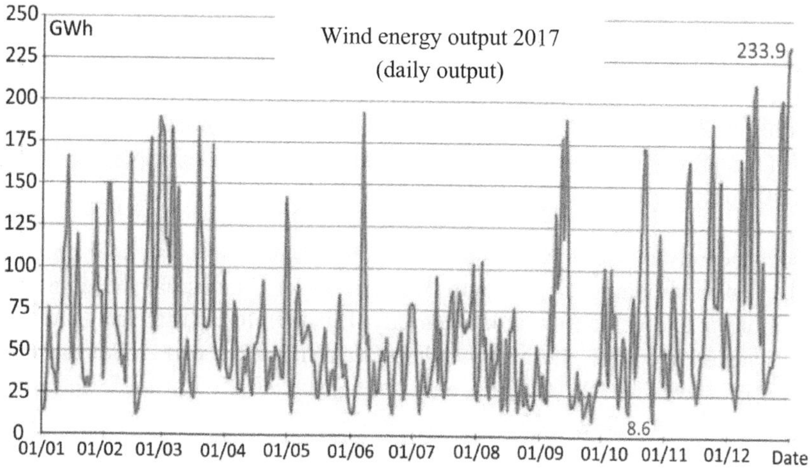

Figure 3.2. *France: wind energy output 2017 (source: RTE – Hulot)*

Though electricity cannot be directly stored, it can be transformed into a reversible potential or into an energy vector, allowing its release in time, or it can be transformed for another storable application.

The most widely known and applied reversible transformations are PHES (pumped hydroelectric energy storage) and electrochemical batteries. PHES pump water from a lower reservoir to an upper one when the electric production is in excess and return it by turbine when needed. These are the most significant installations known, but they require high flow rates and large level differences, which are only available in a limited number of locations. Electrochemical batteries store electricity in a reversible electrochemical reaction. They are convenient to use, are being intensely developed, but also have significant costs, use rare raw materials and have not the potential of PHES. A third method for storing electricity, though less known, uses compressed air (compressed air energy storage (CAES)). The associated heat storage may be a key to its efficiency and will be described in Chapter 4.

Another way to smooth the high variability of renewable generation is to transform electricity into an energy vector, reversibly or not. Hydrogen, and also heat and cold, can be accumulated this way. In France, home hot water tanks that start operating during the night have been the main method for storing nuclear electricity generated during off-peak hours.

Three types of high-temperature heat storage applications for electricity can be identified. One of them is associated with compressed air energy storage to increase its efficiency and reduce greenhouse gas (GHG) emissions. Another one will reversibly transform heat into electricity, which is known as Carnot batteries. Finally, heat storage is directly associated with concentrated solar power plants via the hot fluids they produce.

Independently of these applications, it is important to note that cold generation for the air-conditioning of buildings or for various industrial or commercial applications is a large consumer of electricity. It is therefore particularly interesting to try to generate cold in the periods when electricity is not expensive (e.g. during the night), or available in large amounts (overabundant wind or solar generation), and store it for later use. This may be an advantageous way to store electricity. Phase change materials, essentially ice, are particularly well adapted.

3.1. Heat storage associated with compressed air electricity storage

Compressed air electricity storage consists of compressing air when there is an excess of electricity and then releasing it using a turbine driving an electricity generator when needed. This technique may be related to PHES, but unlike water, air is a compressible gas that warms up when compressed and cools down when expanded, following well-known laws of thermodynamics. The concept is partly complementary to PHES, in the sense that no large height differences are needed, and air is available everywhere. It however requires the use of large underground cavities if significant storage is expected.

The first CAES in the world was built at Huntorf in Germany (Crotogino et al. 2001; Chen et al. 2016). Air is stored under pressure in two large salt cavities of over 150,000 m^3 each, located in depth. Compressors compress the air at 70 bars when electricity is in excess and then it is turbinated at atmospheric pressure, driving an alternator, to return the stored electric energy. The difficulty comes from the fact that direct compression of the air generates heat that may reach 700°C if compression was performed in one step. Besides the fact that this is beyond the capacity of standard compressors, the salt cavity would melt down. The air is therefore compressed by compressors in series and cooled down after each stage, which leads to a loss of energy, and therefore of efficiency.

Furthermore, according to thermodynamics, during the release/restitution, turbining air at 70 bars and moderate or ambient temperature would lead to very low outlet temperatures at atmospheric pressure, incompatible with a proper operation of the system. Therefore, the air has to be heated with natural gas before turbining, reducing the efficiency and emitting greenhouse effect gases, in amounts that are far lower compared to those resulting from a simple gas turbine or a coal-fired power plant use. The final efficiency of the installation in Huntorf is of the order of 45%, compared to over 70% for a PHES, or 90% for a lithium battery, for example. Operational pressure in the salt cavern ranges between 40 and 70 bars, allowing good turbining efficiency.

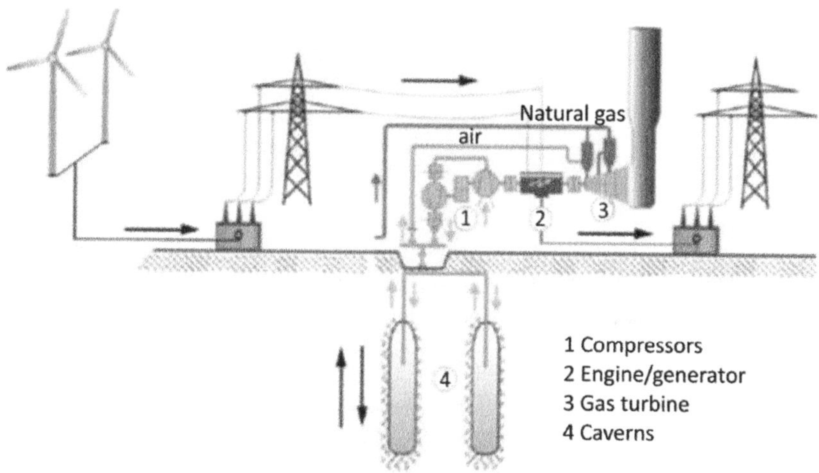

Figure 3.3. *CAES Installation in Huntorf (source: Crotogino et al. 2001)*

The second CAES in the world, built at McIntosh in Alabama, recycles at the turbine inlet a part of the heat of burnt gases and its efficiency is thus above 50% (Chen et al. 2016).

3.1.1. *Adiabatic CAES*

The idea would be to create adiabatic compressed air energy storage (ACAES), avoiding the loss of heat of the air at the compressor outlet and the associated loss of efficiency. Heat would then be stored in one or several stages in specially designed exchangers (thermal energy storage (TES)). They are already widely used in the glass industries, where they are known as regenerators. Multiplying TES allows the reduction of intermediate temperatures, which is in favor of compressors and turbines, but increases the volumes to be implemented.

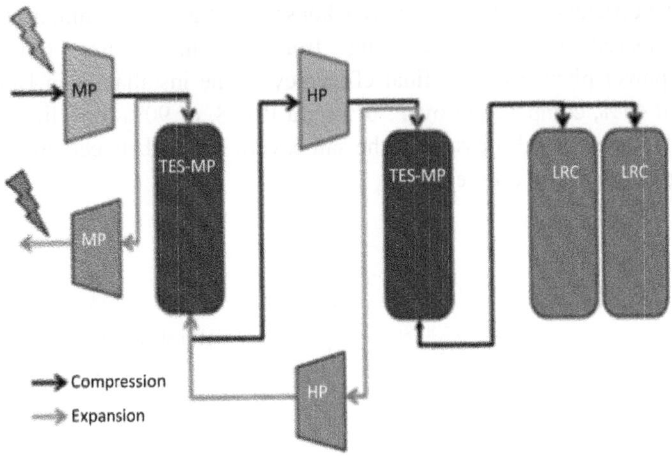

Figure 3.4. *Example of ACAES with two compression stages (source: Nadau 2015). For a color version of this figure, see www.iste.co.uk/odru/heat1.zip*

COMMENT ON FIGURE 3.4.– *TES MP: Thermal energy storage medium pressure (MP). Red capacity on the left: TES HP: Thermal energy storage high pressure (HP). Blue capacities on the right: LRC: Lined rock cavern or compressed air tank.*

After the first compressor, the hot air compressed at an intermediate pressure penetrates into the low pressure (LP) heat exchanger, which is at ambient temperature. It transfers its heat and exits cooled. It is then compressed a second time, once again heated, and then penetrates into the second heat exchanger at HP, where it transfers once again its heat and exits at moderate temperature. It can then be stored in the cavity. Release is done symmetrically: the air stored at moderate temperature and HP is first heated in the HT exchanger before being turbinated a first time and exits cooled. It is then reheated in the LP exchanger before being turbinated a second time and released in the atmosphere.

Considering the hypothesis of a perfect gas, the equations allowing the calculation of constants P, V, T and the internal energy U are well known ($\gamma = Cp/Cv$, Cp: heat capacity at constant pressure, Cv: heat capacity at constant volume):

$$P_1V_1^{\gamma} = P_2V_2^{\gamma} = \text{Const} \qquad PV = nRT$$

$$T_2 = T_1\,(P_2/P_1)^{(\gamma\text{-}1/\gamma)} \qquad U = mC_vT$$

The ACAES thus conceived would no longer require the use of natural gas and its theoretical efficiency would reach 70%. The technological difficulty beyond compressors and turbines is heat storage.

A first attempt at building an ACAES took place in Germany with the ADELE project (Zunft 2015) designed to generate 85 MW power. The technological limitations identified are the compressors and turbines, which must be designed to withstand HPs and high temperatures at inlet and outlet (100 bars and 600°C), far higher than those commonly encountered and used in the industry; then, the design of the two TESs that must ensure resistance to pressure, thermal insulation and efficient heat exchanges with a material storing heat directly. They are built from two 40-m-high containers with rock or ceramic beds. A coating consisting of a heat insulating layer, followed by an external layer of high strength reinforced concrete, provides heat insulation. Compressed air is stored in the salt cavities at a pressure of maximum 65 bars. For both technical and economic reasons, this industrial project does not appear to have been successfully achieved.

3.1.2. *Thermal energy storage*

In the studied cases, thermal storage is achieved in solid materials through which flows the air pressurized by a compressor. Initially cold, these materials must store the heat of the air flowing through them after compression and be able to release it later. TES must therefore be thermally dimensioned to this effect by the heat capacity of the materials used and their exchange capacity under the air flow, while ensuring thermal insulation and resisting pressure. The whole system must have a long lifetime under thermal and mechanical stresses, but also under the effects of corrosion induced by the water and pollutants present in the large amounts of sucked air, and all this should be environmentally friendly and cost-effective. A number of research projects have been conducted in France on these subjects.

3.1.3. *Search project*

This is a research and development project (Puchala et al. 2009; Nadau et al. 2015) aimed at developing a system for the storage of adiabatic compressed air, conducted by ENGIE, Saint-Gobain, CEA Liten and Armines and supported by ANR (French National Research Agency). The compression power is 500 MW during a maximal 5-h period. The expansion power is 200 MW during 8 h, thus releasing 1.6 GWh energy.

Rotating machines are available and industrialized. Classic engineering works are needed in order to adapt the technology to the project but no R&D development is foreseeable. The difficulty comes from the lack of market visibility for these storage conditions, independent from technologies.

On the other hand, the control of the components assembly (machines and process start time, downtime, compression and expansion inversion time, etc.) and the lack of technologies for the thermal storage that may meet the technical installation criteria indicate the need to develop R&D and demonstrators at scale 1.

The preliminary work conducted in the project has been both theoretical and practical. Heat exchangers, which are two, are caverns dug in a metamorphic rock having an inside liner. Expected temperatures are 600°C, with an operating pressure that may reach 150 bars. Numerical simulations have been used to define the overall geometry of the system.

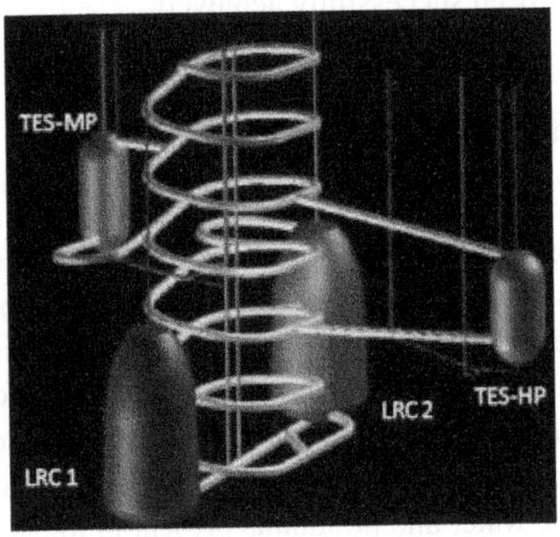

Figure 3.5. *Layout of caverns and their communications. For a color version of this figure, see www.iste.co.uk/odru/heat1.zip*

The two TESs (MP and HP) are cylindrical, being charged by the upper part and discharged by the lower part. They are connected with the two compressed air tanks (LRC1 and LRC2), a series of galleries allowing communications. Mining diggings were validated, with a small hazard on the rock on a favorable site, but an uncertainty on the size related to the necessity of a sealing liner. According to CFD simulations, the system is thermally stable at the fifth cycle, with very small losses (0.5–1%). The two major components studied are the thermal storage materials and their assembly, as well as the sealing liner.

3.1.3.1. *Storage materials*

Storage materials must absorb the heat flow resulting from the air compression and release it during expansion. Therefore, they must have high heat capacity and conductivity, large stability to mechanical and thermal stresses, and also be resistant to corrosion as the sucked air contains water and pollutants.

Ceramic materials prove to be better adapted than natural materials. They have better thermomechanical capacities, allowing reduction of volume, and their resistance in time is satisfactory. In the form of bricks, they allow vacuum optimization and thus avoid the mechanical effects on the walls due to the compression of balls during the cycles (Jansen effect). A high heat capacity iron oxide-based material was chosen ($\rho\ C_p = 2{,}700$ kJ/kg/m^3).

Resistance to corrosion has retained special attention: the air is charged with water vapor which, according to the pressure and temperature conditions, will condense. A thermodynamic model of its behavior under these very special conditions had to be established. But the air is also charged with NOx and SOx, which are assumed to be dissolved in condensed water. A pH of penalizing value 3.5 can thus be deduced. Tests and modeling lead to the conclusion of a proper resistance of the ceramics cycled under these conditions for 30 years.

Structured ceramics are designed to leave a low vacuum rate, below 20%, which optimizes heat capacity (and therefore the stored energy), while allowing air to flow and exchange heat. Simulations have been conducted using a one-dimensional model coupling the air flux and thermal exchanges on a 1,000 m^3 tank storing 930 MJ. The chosen cycle varies between 100 and 600°C at 30 bars 4 h of charge and 6 h of discharge. A 94% efficiency can thus be obtained with high level iron oxide structured ceramics.

It is important to note in this field that CFD simulations on the thermal behavior of a first experimental ACAES built in a tunnel have been successfully conducted (Zavatoni et al. 2019).

3.1.3.2. *Tests conducted on the CLAIRE loop of CEA*

CEA Grenoble manages a dedicated test loop (Figure 3.6) comprising two volumes of 5 m^3, operating at atmospheric pressure. The temperature reached in this test is 600°C. A burner and a system of valves allow several flows of hot or cold air. Balls or structured ceramics were subjected to four air flow rates (between 0.2 and 0.8 kg/s), and pressures and temperatures were measured by a large number of sensors. A mass flow rate ranging between 0.4 and 0.5 kg/s seems optimal as it allows a better exchange between the hot air and the material and therefore a larger energy storage in the volume. The material in the form of balls seems slightly more efficient, due to larger exchange surfaces, but the mechanical stresses induced on the structure by the repeated volume modifications during operation would be highly penalizing.

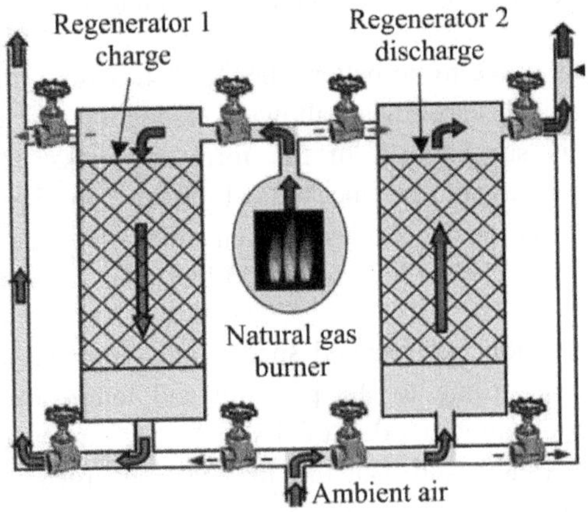

Figure 3.6. *CLAIRE loop of CEA. For a color version of this figure, see www.iste.co.uk/odru/heat1.zip*

3.1.3.3. *Cavern liner*

The cavern liner must provide sealing in time despite repeated thermal and mechanical stresses, occupying a minimal volume.

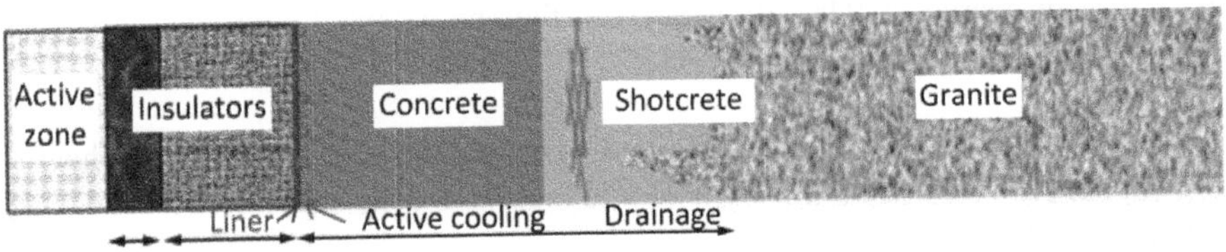

Figure 3.7. *Liner inside the cavern. For a color version of this figure, see www.iste.co.uk/odru/heat1.zip*

The proposed optimized liner is shown in Figure 3.7. Leaning against the hosting rock, it comprises two concrete layers, then an actively cooled metallic liner, and two layers of thermal insulators in contact with the active zone. The operating structure undergoes strong thermal exchanges, particularly through convection, and also pressure efforts, inducing stresses in pressure and compression in temperature, with the modification of the characteristics of the materials. A complex model coupling thermal, mechanical and hydraulic aspects for deformable porous media saturated with a compressible fluid had to be developed in order to dimension it. The efficiency of the liner is such that the hosting rock does not participate in the heat storage.

3.1.3.4. *Economy*

Coated or not, the mined cavities do not seem to be economically justified when their resulting levelized cost of storage (LCOS) is compared to that of PHES. This must be put into perspective by a complementary geological typology and a smaller environmental footprint. However, this approach of ACAES in mined cavity remains competitive compared to other electrochemical storage technologies for such levels of stored power and energy. Surface or subsurface TES associated with saline caverns should probably be less expensive.

As for all electricity storage, *peak-off peak* arbitrage is not sufficient for obtaining an efficient ACAES. The business model must be combined with other grid services (congestion, shift of investment, tertiary reserve, capacity market, etc.) and long-term prices are guaranteed by a political framework. Recently, signals in favor of ACAES storage can be perceived throughout the world, particularly for the technology developed by Hydrostor[1].

3.1.4. *SACRE project*

A second project was conducted in France on this subject by École Polytechnique, PROMES, L2EP Lille, EDF and Geostock (Berest 2010). According to this project, ceramics resulting from the processing of asbestos wastes will be used as thermal storage materials. In order to be rendered harmless, the asbestos once used for insulating buildings is subjected to a high temperature process that transforms it into a ceramics with very good thermal properties for being used as heat storage, with the advantage of being very cheap, as it is considered a waste. Another material with similar characteristics results from fly ashes from the steel industry (Table 3.1).

Tests conducted at 30 bars and 600°C on asbestos ceramics indicated no evolution (diffractogram, mechanical properties) after 250 h, then up to 750 h. It is important to note that these materials have been initially proposed and characterized for use in heat storage associated with concentrated solar power plants.

1. Available at: www.hydrostor.ca/projects.

Material	HT concrete	Molten salts	HT ceramics	Asbestos waste ceramics	Fly ashes ceramics
ρ (kg m^{-3})	2,750	900–2,600	3,500	3,120	2,975
Cp (J kg^{-1} K^{-1})	915	1,500	870	800–1,035	715–1,220
ρ Cp (kJ m^{-3} K^{-1})	2,520	1,350–3,900	3,030	2,500–3,200	2,120–3,340
λ (W m^{-1} K^{-1})	1	0.15–0.2	1.35	2.1–1.4	1.16–1.59
CTE (10^{-6} K^{-1})	9.3	–	11.8	8.8	10–1,200
Cost (€/tonne)	80	750	4,500–9,000	8–10	10–1,200

Table 3.1. *Characteristics measured for a range between ambient temperature and 1,000°C of various materials for high temperature storage, ρ: density, Cp: specific heat capacity, λ: thermal conductivity, CTE: thermal expansion coefficient*

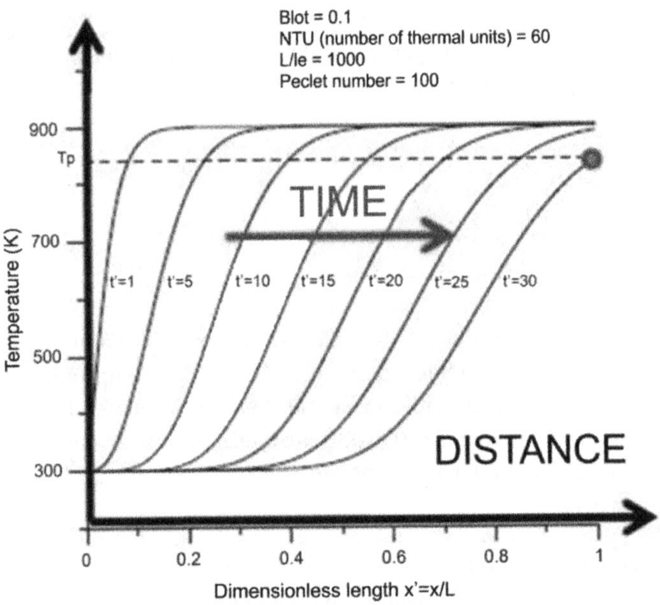

Figure 3.8. *Progression of the thermal front in TES and breakthrough time*

The system was solved depending on various dimensionless numbers, allowing the description of the progression of the thermal front in TES and also of the breakthrough time (Figure 3.8).

The efficiency is optimum for a small difference between minimal and maximal pressures in the cavity, which explains the necessity of large cavities. ACAES must be located as close as possible to the need, and its economy will depend on the availability of adapted caverns. Pressure variations accompanying the system operation induce variations of mechanical and thermal stresses, leading to cyclic fatigue that must be taken into account in engineering.

3.2. Electricity storage by Carnot batteries

As indicated in the introduction, electricity is not stored directly, but it can be stored indirectly by transformation into a reversible potential. This chapter is dedicated to a process for storing electricity in the form of high temperature heat, and Table 3.2 illustrates its potential.

Water at 100 m	Li-ion battery	Concrete at 800°C
275 Wh	220 kWh	217 kWh

Table 3.2. *Energy stored by tonne in the given conditions*

The strong potential of heat can be noted. However, the reversible transformation into electricity is not intuitive: for example, heat extraction from concrete leads to its progressive decrease and to a rapid loss of interest.

The idea is therefore to find a process for the storage of electricity in the form of high-temperature heat allowing for constant release with high efficiency. A process of this type was proposed and patterned by the Saipem SA company in the form of a modified Brayton cycle (Ruer 2016).

3.2.1. *Modified Brayton cycle: operating principle*

The process will use a gas as heat transportation vector. It is composed of two identical chambers, an LP chamber and a HP chamber connected by two systems of coaxial compressors and turbines, a high temperature one and a low temperature one. Each of the chambers is filled with a combination of optimized heat storage materials. There is a cold point between the HP storage and the turbine and the coaxial cold compressor.

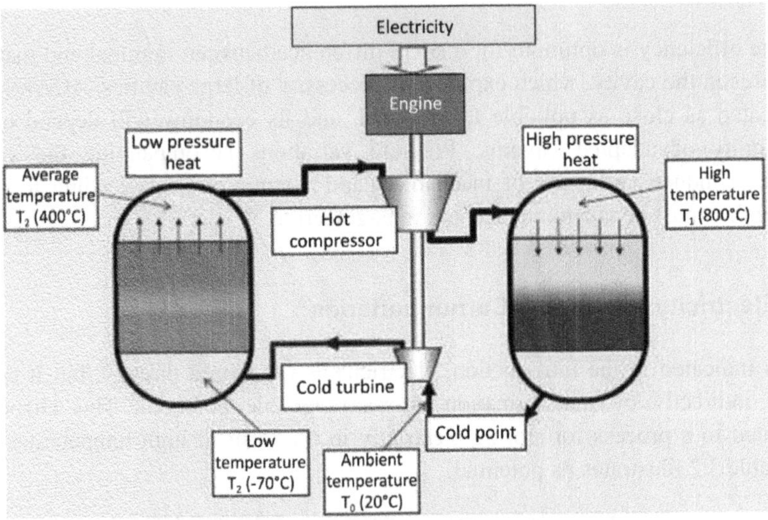

Figure 3.9. *Modified Brayton cycle during the storage phase (source: Saipem). For a color version of this figure, see www.iste.co.uk/odru/heat1.zip*

In the chosen example, in the initial conditions, the LP chamber is at a temperature of 400°C and the HP one at ambient temperature of 25°C. Excess electricity is used to suck the gas via a compressor from the chamber at 400°C to the chamber at 25°C. The compression heats the gas that enters the HP chamber at 800°C. For a perfect gas, the compression rate would be of the order of 5 for nitrogen, but for argon, a monoatomic gas (adiabatic exponent $\gamma = 1.67$), it would be only 3.2, explaining the preferred choice of this gas.

In the HP chamber, the gas at ambient temperature is pushed and feeds the cold turbine from which it exits at LP at a temperature of –75°C. The expansion energy of the turbine is transferred to the compressor shaft and increases the efficiency of the cycle.

During the discharge, the system is reversed. The hot air at 800°C is sucked from the HP chamber to the LP chamber via the high temperature turbine. Its expansion transfers energy to an alternator, producing electricity. It goes out at 400°C. The cold air of the LP chamber is sucked by the cold compressor, and goes into the HP chamber at ambient temperature after passage at the cold point. It should be noted that the hot turbine supplies the required energy to the cold compressor, thus reducing the efficiency of the system.

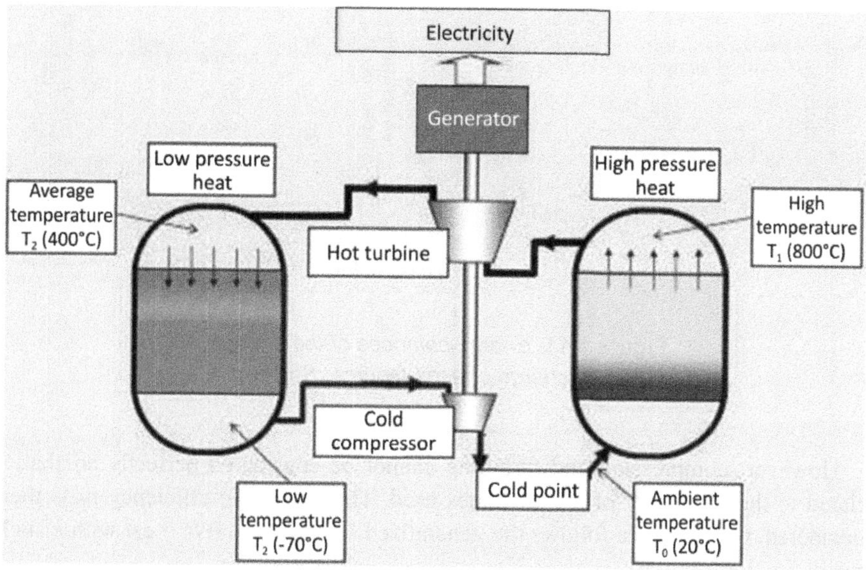

Figure 3.10. *Modified Brayton cycle in discharge phase (source: Saipem). For a color version of this figure, see www.iste.co.uk/odru/heat1.zip*

3.2.2. *Modified Brayton cycle: energy balance*

The energy balance is written by considering that the internal energy of a gas is proportional to its absolute temperature. For a perfect gas, the equations are well known:

$$P_1V_1^{\gamma} = P_2V_2^{\gamma} = Cst \qquad PV = nRT$$

$$T_2 = T_1 (P_2/P_1)^{(\gamma-1/\gamma)} \qquad U = mC_vT$$

It can be readily noted that if the system was ideally reversible (efficiency of all the unit elements such as compressors, turbines, but also heat transfers in the chambers, thermal fronts and absence of losses), the efficiency would be equal to 1.

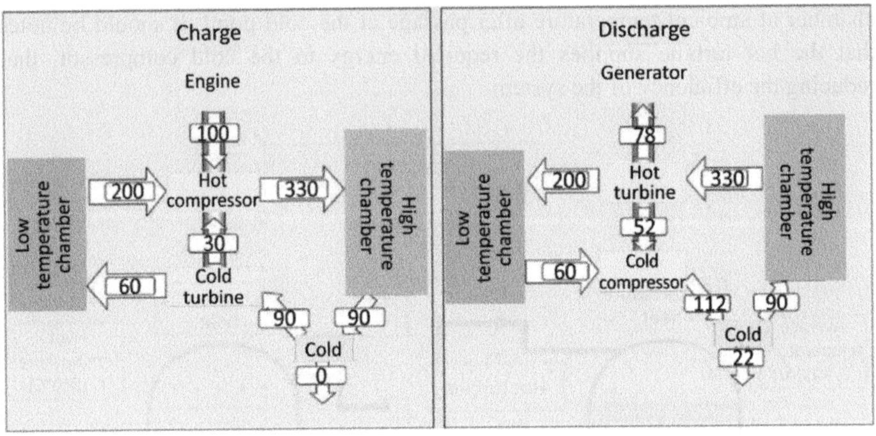

Figure 3.11. *Energy balances of the storage and discharge phases (source: Saipem)*

However, compression and turbining cannot be considered perfectly adiabatic, related to the efficiency of the machines used. The polytropic efficiency η_p is then considered. Gas pressure follows the generalized Laplace law PV^k = cst with k such that:

$$\frac{k-1}{k} = \frac{\gamma - 1}{\gamma}\frac{1}{\eta_p}$$

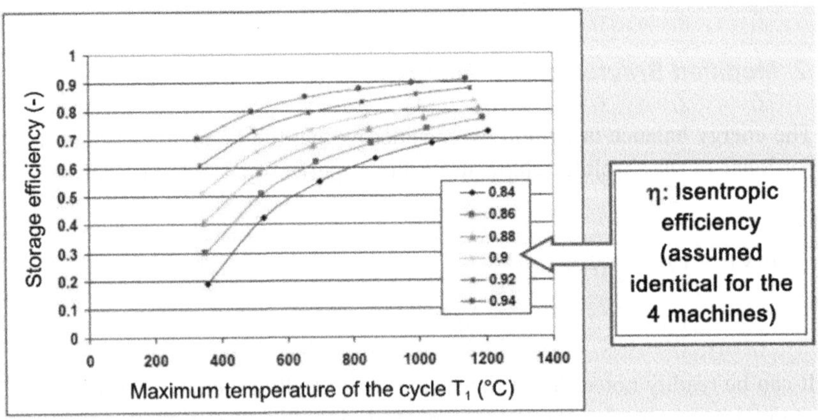

Figure 3.12. *Storage efficiency depending on the maximal temperature and on the isentropic efficiency of rotating machines (source: Ruer 2016). For a color version of this figure, see www.iste.co.uk/odru/heat1.zip*

The outlet temperature will be higher than in case of a perfect transformation. It can be deduced that the irreversibility related to thermal machines will be reflected by an increase in the temperature of the real system during time. This heat should therefore be evacuated to the external environment, thus justifying the cold point between the HP tank and the expansion turbine.

It can be noted that the same evacuation could also be added between the LP tank and the hot compressor, but a high-temperature exchange is more complex to achieve. In these conditions, the efficiency of the system was estimated for various values of the isentropic efficiency assumed identical for the four machines.

It can be noted that the efficiency of the storage is sensitive to the isentropic efficiency and increases with temperature. It is of the order of 70% toward 800°C for a good efficiency of the machines. A higher temperature would lead to better values, provided that adapted compressors are available.

3.2.3. *Feasibility*

As for any innovative project, the key elements and the potential technological difficulties have to be identified. Through a research project (Patenotre 2008) conducted by Saipem, Poweo, CEA Grenoble, ONERA, CNAM, ARMINES and ENSCIL, and supported by ANR, and a thesis (Desrues 2011), several points have been studied.

3.2.3.1. *Availability of rotating machines*

According to the study of rotating machines, the materials for working at 800°C are known, and the rotation speed is sufficiently low to avoid centrifugal effects. Cooling the blades is not necessary. The clean and neutral gas flowing through a closed circuit eliminates the corrosion risk.

There are turbines operating under these conditions. Not the same is true for compressors, and not for genuine feasibility reasons, but for practical ones. Indeed, in the industry, gases are compressed through series of compressors, and cooled between each of them, so that unlike turbines, hot compressors are not available. Though their development is not a technological difficulty, it would be lengthy and costly, requiring a market that does not currently exist.

In these conditions,.an isentropic efficiency of 0.88 may be reached.

3.2.3.2. *Storage materials*

The heat storage material must have some characteristics: good mechanical strength to heat, resistance to cyclic thermal fatigue, high heat capacity, good thermal conductivity and low cost. Its layout must also reduce charge losses and avoid the Janssen effect, which is caused by the cycling of spherical materials in a cylinder, and is reflected by strong stress that may lead to burst. Natural materials, such as olivine basalt, inexpensive, could be considered, but must be calibrated and cleaned. Ceramics (cordierites) have the advantage that they can be shaped to meet the specifications.

3.2.3.3. *Thermal exchanges*

As for the study of the ACAES system, an arrangement of storage materials should be defined so that it optimizes the amount of stored heat and the thermal exchanges during the gas circulation, while minimizing charge losses, particularly in the cold tank, where the circulation is more rapid and the pressure is low. According to the research results (Ruer 2016), these objectives can be reasonably met. Various geometries have been tested. The quantities of materials and the volume of the necessary tanks can be deduced. Spheres present advantages, but they induce the Janssen effect. The use of ceramics with turbulence promoters (alternate obstacles) intensifies exchanges without increasing too much the charge losses (Desrues 2011; Desrues et al. 2012).

Thermal fronts circulate at very low speed and are not perfect. According to the study, they tend to get homogeneous after a number of cycles, but need to be adjusted.

Finally, the entire system should have a very efficient thermal insulation. These points contribute to a reduction of the overall efficiency of the system.

3.2.3.4. *Balance*

The numerous theoretical and experimental studies conducted (some of them on the CLAIRE loop of CEA, Figure 3.6) allowed the validation of the following points:

– thermodynamic feasibility of the process. Complete modeling was achieved;

– feasibility of turbo machinery, including the hot compressor, which is however not available on the market;

– validated thermal storage materials;

– feasibility of chambers and optimization of thermal exchanges and charge losses.

3.2.4. *Potential characteristics of a thermal pumping system*

The potential characteristics of a thermal pumping system have been established by Saipem using the example of a 10 GWh storage:

Power	100 MW
Dimensions of the tanks – Diameter – Height	 65 m 45 m
Hot tank – Pressure – Maximal temperature	 4 bars relative 800°C
Cold tank – Pressure – Minimal temperature	 Ambient 70°C
Mass of basalt gravel – Efficiency – Self-discharge	250,000 t 70% 0.5%/day

Table 3.3. *Characteristics of a thermal pumping system (source: Saipem)*

Figure 3.13. *Artist view of a 10 GWh thermal pumping system (source: Saipem). For a color version of this figure, see www.iste.co.uk/odru/heat1.zip*

The potential advantages of a thermal pumping system can be summarized as follows:

– electricity storage of medium to large capacity 100 MWh–10 GWh (even more);

– charge/discharge time of several hours (to several days);

– acceptable efficiency (0.70);

– limited pressure (4–5 bars);

– maximal conventional temperature (800°C);

– energy density: 30–50 kWh/m^3;

– absence of dangerous products;

– no site constraint.

3.2.5. *Brayton cycle variant*

A modified Brayton cycle variant was proposed by Farres-Anthuniez (2018). The main difference is that compression or expansion heat is transferred through exchangers to liquids that operate at atmospheric pressure, according to Figure 3.14. The four temperatures are stored in four distinct tanks instead of two, and there is no thermal front.

In storage phase, excess electricity activates the compressor that heats the gas sucked at intermediate temperature after passage through the cold exchanger. This hot gas exchanges its heat in the hot exchanger with the fluid at intermediate temperature, which is then stored in the hot tank. The gas exits at intermediate temperature and is expanded in the turbine, then goes out of it cooled. Cold is transferred to the fluid and then stored in the cold tank. The turbine expansion reinforces the pumping and increases the efficiency of the system. During discharge, hot air expands in the hot turbine and activates the cold compressor, driving the reverse circulation of fluids and heat exchanges. Heat exchangers with the external environment allow the evacuation of irreversibilities.

This system features a number of differences compared to the previously described system, in which the gas exchanges its heat with solid materials. The fluid storage chambers are independent and therefore they are not pressurized. The pressure of the gas circuit can thus be increased, rendering the system more compact. The thermal exchange front is suppressed. However, the system becomes sensitive to the efficiency of heat exchangers. Finally, the operational temperature ranges of the liquids are narrower than those of solids, which leads to temperature limitations. This may be compensated by the use of two of them in series (typical example, mineral oil, then molten salts), for example.

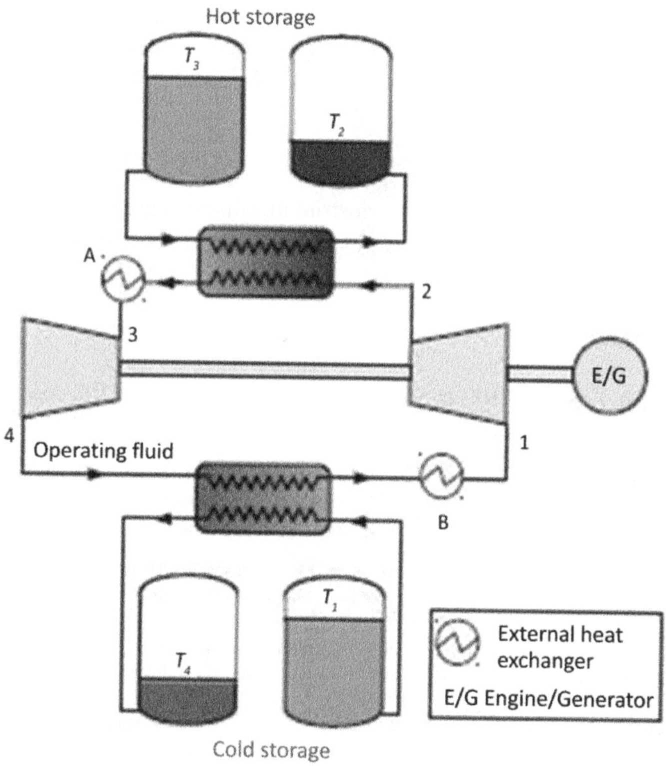

Figure 3.14. *Brayton cycle variant with liquid exchangers (source: Farres-Anthuniez 2018). For a color version of this figure, see www.iste.co.uk/odru/heat1.zip*

The authors obtain an efficiency that may reach 70% and costs that may be equivalent to those of the hydraulic pumping storage systems, provided that the rotating machines and exchangers reach significant efficiencies.

3.2.6. *Phase change cycle*

Cycles can be imagined including a phase change. The most accomplished one is LAES (liquid air energy storage) (see Ruer 2016; Farres-Anthuniez 2018). After preprocessing, the air passes in a compressor train activated by excess electricity. It is cooled at each stage, and the corresponding heat can be stored. The pressurized air at nearly ambient temperature is then expanded and transformed into cold liquid, which is then stored.

During the discharge operation, liquid air is pumped at HP, heated and vaporized, then expanded in a turbine system driving a generator. The recovery of a part of the generated cold and the use of a heat source would allow reaching efficiencies of about 70%.

The interest in such a system is given by the fact that all the components are industrially available. Pilot and even industrial installations are under construction in the United States and Great Britain (Highview Power Storage 2017).

3.3. References

Berest, P. (2010). Projet SACRE. In *Séminaire mi-parcours Stock-E 2010*. Agence Nationale de la Recherche, Paris.

Chen, L., Zheng, T., Mei, S., Xue, X., Liu, B., Lu, Q. (2016). Review and prospect of compressed air energy storage system. *Journal of Modern Power Systems and Clean Energy*, 4, 529–541.

Crotogino, F., Mohmeyer, K., Scharf, R. (2001). Huntorf CAES: More than 20 years of successful operation. In *SMRI Spring Meeting*, Orlando, Florida.

Desrues, T. (2011). Stockage massif de l'électricité sous forme thermique. PhD Thesis, Université de Grenoble, Grenoble.

Desrues, T., Ruer, J., Marty, P., Fourmigué, J.F. (2010). A thermal energy storage process for large scale electric applications. *Applied Thermal Engineering*, 30(5), 425–432.

Desrues, T., Marty, P., Fourmigué, J.F. (2012). Numerical prediction of heat transfer and pressure drop in three-dimensional channels with alternated opposed ribs. *Applied Thermal Engineering*, 45/46, 52–63.

Farres-Anthuniez, P. (2018). *Modelling and development of thermo-mechanical energy storage*. PhD Thesis, University of Cambridge, Cambridge.

Highview Power Storage (2017). Liquid air energy storage. Pumped hydro capability. No geographical constraints [Online]. Available at: www.highviewpower.com/wp-content/uploads/2018/04/Highview-Brochure-November-2017-Online-A4-web.pdf.

Laughlin, R. (2017). Pumped thermal grid storage with heat exchange. *Journal of Renewable and Sustainable Energy*, 9.

Nadau, L., Leplay, P., Fourmigue, J.F., Memponteil, A., Hadj Hassen, F. (2015). A regenerator pilot to evaluate the technical and economic relevance of energy storage by adiabatic compressed air energy storage by ceramic media. In *9th International Renewable Energy Storage Conference (IRES 2015)*.

Odru, P. (2013). *Le stockage de l'energie*. Dunod, Malakoff.

Patenotre, A. (2008). Appel à projets Stock-E 2008 – Projet SETHER. Agence nationale de la recherche.

Puchala, E. (2009). Projet SEARCH. In *Séminaire mi-parcours Stock-E 2009*. Agence nationale de la recherche.

RTE (n.d.). La consommation d'électricité en France [Online]. Available at: www.rte-france.com/eco2mix/la-consommation-delectricite-en-france.

Ruer, J. (2016). Climat Énergie Océan. *ASPROM*.

Zavatoni, S.A., Geissbühler, L., Zanganeh, G., Andreas, H., Aldo, S., Barbato, M.C. (2019). CFD modeling and experimental validation of the TES unit integrated into the world's first underground AA-CAES pilot plant. In *AIP Conference Proceedings*.

Zunft, S. (2015). The ADELE ING project. In *SCCER Heat & Electricity Storage Symposium*. PSI, Villigen.

4

Latent Heat Storage: Fundamentals and Most Widely Used Phase Change Materials

Ana LAZARO[1] and Erwin FRANQUET[2]

[1] *GITSE, I3A, University of Zaragoza, Spain*

[2] *Polytech'Lab, Université Côte D'Azur, Nice, France*

4.1. Fundamentals of latent heat storage

Though many people are not so familiar with the concept, latent heat storage has been used for decades, let alone millennia. As a matter of fact, the main purpose for us was to make reserves of food to secure the amount of energy we will require for our basic needs (e.g. to maintain our internal temperature, to have our organs working properly, etc.) and then, later on, to permit us to save time for other activities. Similarly, the wooden logs at the back of the house and the fuel put in the tank are energy stored for a future use; wooden logs will be burned to provide heat, and the fuel in the tank will be used instead of human labor to move the car (and so therefore people). This being said, it is worth mentioning that energy can rarely be stored in its final useful form but, in contrast, it can be stored in many different ways, for instance, under mechanical, potential, chemical form or heat. The aim of the present contribution is to focus on thermal energy storage, and more especially on latent heat storage (LHS), so ignoring sensible heat storage and thermochemical storage. For more details on energy storage in general, readers can look at Huggins (2010) and Sterner and Stadler (2019), and for the specific case of heat storage, refer to Dincer and Rosen (2002) and Paksoy (2007).

Heat and Cold Storage 1,
coordinated by Pierre ODRU, Elena PALOMO DEL BARRIO.

To recall the main principles of thermal energy storage, one has to briefly detour by thermodynamics. Being schematically the science of the "movement of heat", it is indeed the branch governing the study of energy exchanges between bodies (involving heat or work or any other form of energy, and their possible transformation into each other). In sensible storage, the energy stored is directly proportional to the temperature increase (respectively decrease) and to the so-called heat capacity of the involved body. For thermochemical storage, the energy storage relies on the exothermic or endothermic needs of some chemical reactions. Lastly, latent storage relies on the amount of energy, according to the first law of thermodynamics, that is needed by a system to transform from one state to another (see Figure 4.1). Depending on the type of transformation, this quantity is referred as latent heat (phase change), enthalpy of transition (crystal structure transition) and heat of dissolution (salt dissolving, i.e., melting in a solvent). It is usually represented as a variation of the specific enthalpy Δh [kJ/kg], with $h = u + P \cdot v$.

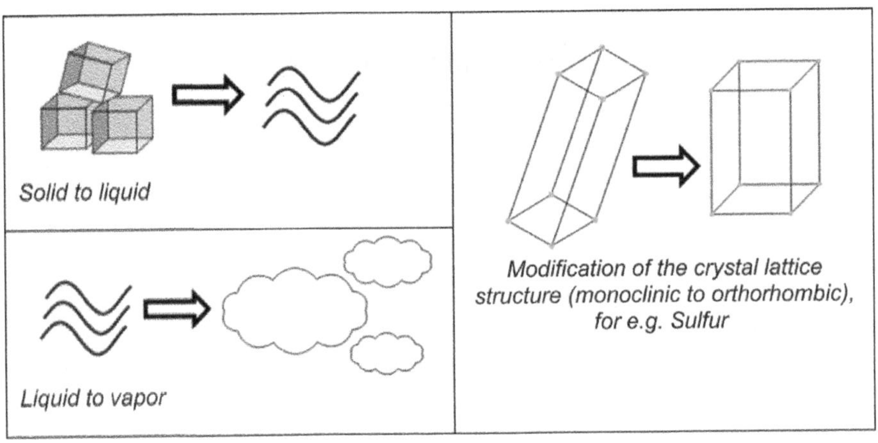

Figure 4.1. *Examples of transformations involving an enthalpy change*

Though formally "latent heat" only corresponds to a modification of the state of matter, it is in fact very common to extend this notation to all kinds of enthalpy changes, for instance, with clathrate hydrates or gas hydrates, or semi-clathrates formation. On the contrary, although all materials are potentially subject to phase change, the so-called phase change materials (PCMs) generally correspond solely to the ones undergoing solid/liquid or solid/solid transitions. Similarly, LHSs usually refer to thermal energy storages involving PCMs (Garg et al. 1985; Mehling and Cabeza 2008). These shortcuts will be adopted in the following to respect the usual conventions. Historically, this could be explained by the reluctance of engineers to rely on liquid/vapor (vaporization) or solid/vapor (sublimation) phase change to store energy. Indeed, despite higher associated latent heats, the corresponding volume change raises many industrial issues in practice and the vast majority of applications were not based on vaporization/condensation changes.

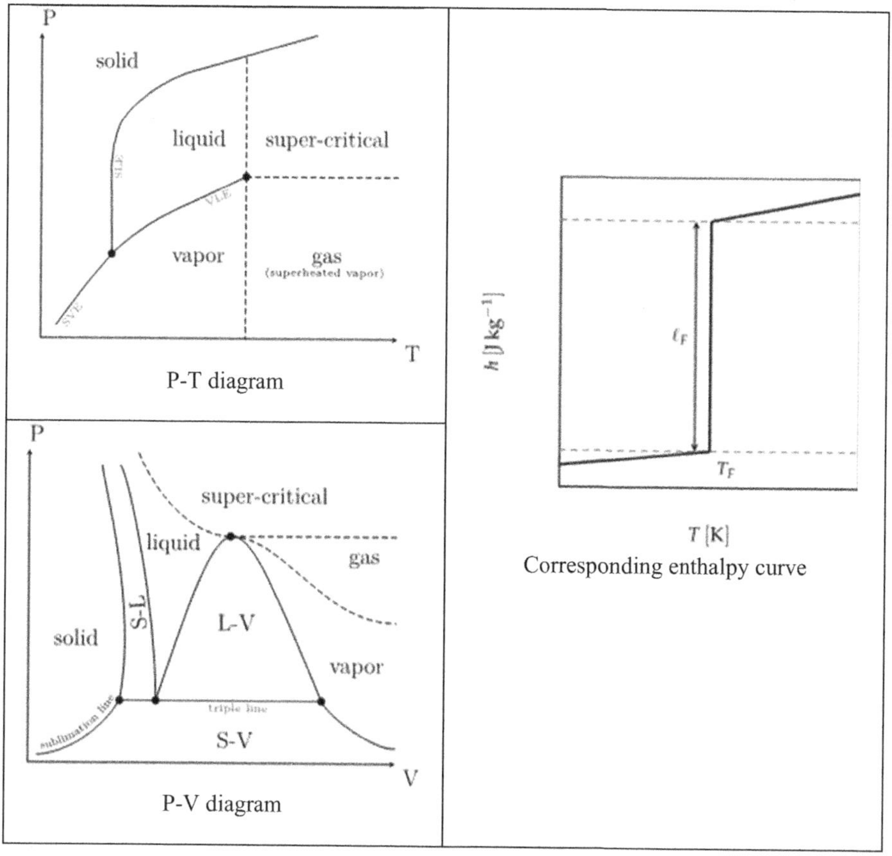

Figure 4.2. *Illustration of a first-order phase change, occurring at constant conditions*

For all the above reasons, liquid/vapor and solid/vapor transitions will not be discussed hereafter. For the remaining possibilities, solid–liquid phase changes present considerably higher latent heats and solid–solid transitions: as an example, the ratio is greater than 30 for sulfur. In the same way, latent heats of fusion/solidification are much larger than what can be achieved with sensible heats, even when considering several tens of degrees of energy stored in the heat capacity.

From a thermodynamic point of view, this amount of energy is either exchanged at constant temperature and pressure, for pure substances or eutectic/peritectic solutions for example (first-order transitions), or on a temperature range that can spread from a few to about tens of degrees, for mixtures and solutions with impurities (second-order transitions). These two different behaviors are illustrated in Figures 4.2 and 4.3. In the first case, temperature is thus fixed during the transition from one state of matter to the other and corresponds to the phase change temperature T_{pc}, which is therefore unique. Besides, this characteristic temperature is specific for the material. For the corresponding enthalpy curve, it is discontinuous at T_{pc}. Figure 4.2(c) presents an example of such a function. Unfortunately, this first situation is rarely encountered since pure substances are usually very expensive or, more precisely, they do not fit the current profitability criteria to withstand the current economic analysis. So, when one turns to the second case, therefore more widespread, the behavior of the material can be singularly more complicated. It is indeed dependent on the mixture classification, which varies according to the interactions and miscibility (or not) between the raw elements composing the mixture and, this, in all phases. A complete presentation being out of scope of the present book, it will not be further detailed here. In brief, the phase change occurs now on a finite domain, associated with a temperature variation. This latter one depends on both the nature of the basic elements, and on their relative concentration in the mixture. Unhappily, the enthalpy curve, which is the main paramount parameter for the present applications, is definitely more complicated to write. An example, of a eutectic solution, is thus shown in Figure 4.3(c): the temperature range has to be defined (in the present example, it would imply explicitly the eutectic and liquidus temperatures) but also the latent heat, which can be the sum of several contributions.

4.2. Phase change materials classification and criteria for selection

Among the different classification criteria that can be applied to a PCM, the most common categorization is based on the chemical nature of the material. Therefore, PCMs are differentiated into organic and inorganic materials. This distinction is used to generalize and to compare the material behavior in general terms of density, enthalpy change and solidification behavior (Khudhair and Farid 2004; Jegadheeswaran and Pohekar 2009; Agyenim et al. 2010; Gil et al. 2010; Kuznik et al. 2011; Ge et al. 2013; Kuravi et al. 2013; Tatsidjodoun et al. 2013; Waqas and Ud Din 2013; Iten and Liu 2014; Yuan et al. 2014; Su et al. 2015; Akeiber et al. 2016). Table 4.1 shows the main chemical compounds that can be used as PCM.

Organic	Inorganic
a) Alkanes (paraffins) b) Alcohols c) Carboxylic Acids (fatty acids) d) Esters e) Ketones f) Amide g) Amine h) Alkene i) Thiol	a) Salt hydrates b) Metals c) Inorganic compounds

Table 4.1. *Latent heat storage material classification attending to their chemical nature*

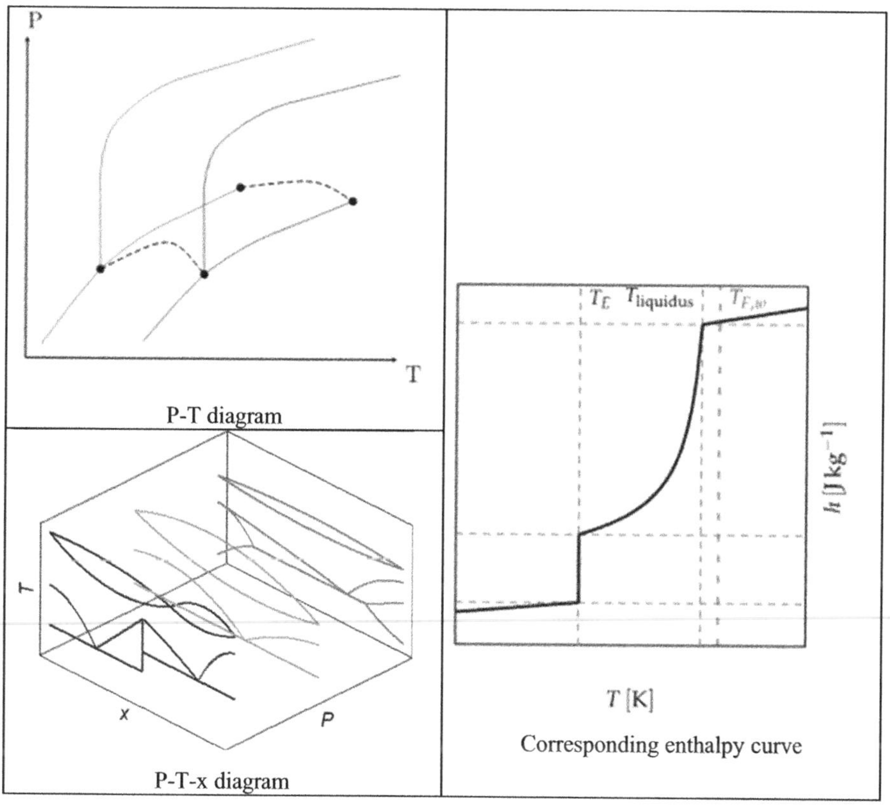

Figure 4.3. *Illustration of a second-order phase change, associated with non-constant conditions (N.B. the sketch presented here is only illustrative and does not correspond to any specific material). For a color version of this figure, see www.iste.co.uk/odru/heat1.zip*

Another classification, regularly encountered, is shown in Figure 4.4. It is based on the composition, which depending on the PCM consists either of pure materials with different grade of purity or of mixtures. Other classifications can be based on the phase temperature, with low temperature for those PCMs with melting temperatures below 10°C, medium temperature from 10°C to 120°C and high temperature over 120°C. Also, the purity grade of the material is a criterion for categorization: ACS (American Chemical Society) grade, RG (Reagent Grade) grade, USP (United States Pharmacopeial Convention) grade, NF (National Formulary) grade, laboratory grade, purified grade and technical grade.

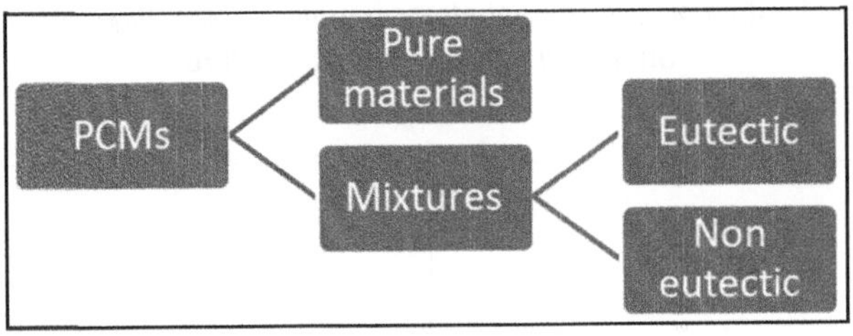

Figure 4.4. *Latent heat storage materials classification attending to their composition*

In spite of these impressive aforementioned advantages, PCMs also present some specific features that can possibly hamper their use and deployment in certain circumstances. In this respect, some features are preferred for materials that awaken significant interest as LHS material. Table 4.2 provides these main desirable characteristics.

Thermal/kinetic	Physical	Chemical	Economic
– Appropriate phase transition temperature – High latent heat of fusion and specific heat – Narrow phase change temperature range – High thermal conductivity – High crystallization rate – No or low supercooling	– High density – Small volume change – High-phase stability – Low vapor pressure	– Compatible with container material – Long-term chemical cycling stability – Highly non-flammable, non-toxic and – Non-explosive – Recyclable	– Cost-effective – Abundant – Commercially available – Wide range application

Table 4.2. *Main desirable characteristics for a latent heat storage material*

No materials can satisfy all the features listed in Table 4.2. However, the phase change temperature, the density of stored energy and the chemical stability are those more relevant to the technical feasibility of any application. Indeed, when a material has to be selected for a storage application, two temperatures are critical: The available heat source temperature and the required heat sink temperature. The heat source temperature corresponds to the temperature of the heat source to be used for charging the heat storage system. The heat sink temperature is then the temperature of the sink of heat during the discharging process. The heat source and the heat sink fluids are usually air or water, but they can be also other fluids like heat transfer process fluids. The PCM will operate between those two temperatures and, consequently, the maximum theoretical storage capacity of the material corresponds to the stored energy between the heat source and the heat sink temperatures.

Furthermore, the heat transfer process relies on a temperature difference, so the minimum temperature of the PCM during the discharging process will be always higher than the sink temperature, and the maximum temperature of the PCM during charging will be always lower than the heat source temperature (Figure 4.5).

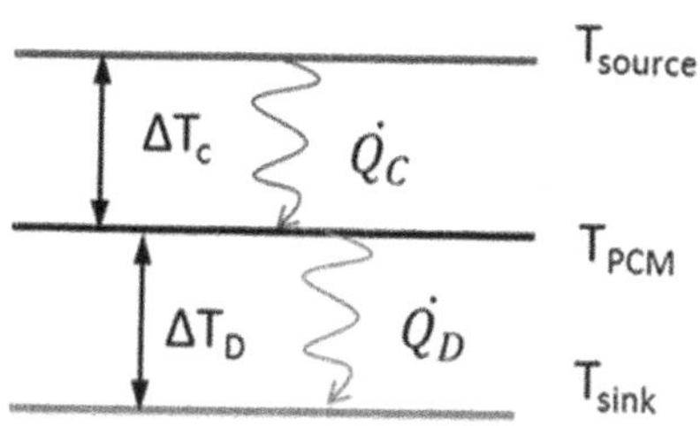

Figure 4.5. *Scheme of temperature levels and its relation to thermal power during charging and discharging for PCM selection in a heat storage application*

With respect to these constraints, the phase change temperature or interval should be selected accordingly to the application temperatures. Moreover, aiming to achieve higher efficiencies in charging and discharging processes, as Carnot efficiency indicates, the temperature difference between the source charging temperature and the phase change temperature or temperature range (ΔT_c) should be as low as possible and, at the same time, the temperature difference between the sink discharging temperature and the phase change temperature or interval (ΔT_D)

should be as low. On the other hand, the lower the temperature difference (ΔT_c or ΔT_D), the higher the heat exchange surface is needed to satisfy the charging/discharging power requirements ($\dot{Q}_C$ and $\dot{Q}_D$). So, as shown in Figures 4.5 and 4.6, there is a compromise when selecting the material, according to the available source temperatures and required sink temperatures, and the thermal power requirement.

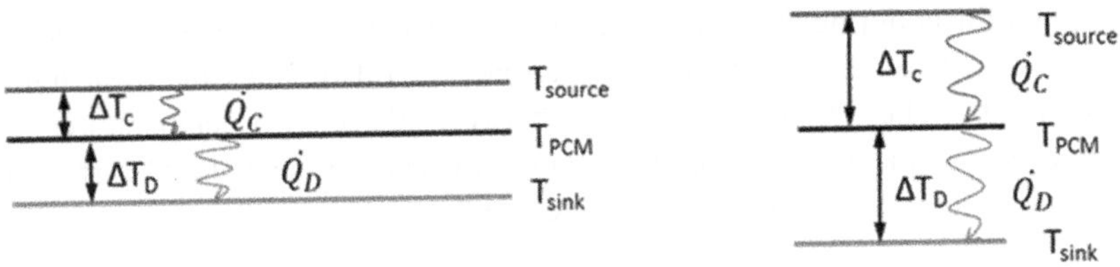

Figure 4.6. *Scheme of temperature levels and its relation to required heat exchange surface and efficiency during charging and discharging for PCM selection in a heat storage application. Left: High-efficiency heat transfer processes, large heat exchange surface required. Right: Low-efficiency heat transfer processes, small heat exchange surface required*

To name but a few caveats, supercooling and incongruent melting of solutions undeniably entail some drawbacks, as well as thermal cycling (Davis 2001; Mullin 2001; Mehling and Cabeza 2008; Glicksman 2011). Not to forget the other issues pertaining with the possible corrosiveness, poisonous nature and/or flammability features, but these ones are not intrinsically tied to PCMs. In fact, it depends on their operation conditions and, consequently, these points will be discussed in the next section, when shedding a light on the criteria governing their selection. Supercooling has been known for centuries now and corresponds to the appearance of metastability in the liquid phase: during a cooling, the material is not solidifying and stays in liquid state while the temperature is below the phase change temperature at which the melting initially took place. In other words, it is observed that the crystallization temperature, representing the start of the solidification process, is lower than the melting temperatures; consequently, and even for pure substances, one has to consider two phase change temperatures. Moreover, it is sometimes required to considerably decrease the temperature (up to 60–80 K) for the crystallization to begin, and the latent heat to be released again by the material. This observation can be explained, for homogeneous nucleation, by the competitive and opposite actions of two driving forces. On the one hand, for a nucleus (that is to say a tiny solid particle) to grow, the system must overcome the surface energy barrier, proportional to the square of the radius, which is required to maintain the nascent solid/liquid interface. On the other hand, the heat gain due to the phase change depends on the volume of liquid crystallizing, and is so proportional to the cube of the radius. If this release of heat is smaller than the previous barrier, then the solidification is hindered. Practically, it can be shown that there exists a critical radius under which the solidification cannot occur, due to thermodynamic instability, and only nuclei with a greater radius can grow. This critical radius decreases with the degree of supercooling DT_{pc}, that is to say with the increase in T_M-T_C and so, the larger DT_{pc}, the higher the probability for the solidification to start. However, this also means that solidification is no more deterministic and can be stochastic, which is rarely appreciated in industrial applications. Added to this, it is also worth mentioning that the degree of supercooling DT_{pc} is not independent of the operating conditions and it tends to increase for small samples and for high rates of cooling. Finally, PCMs with supercooling have a wider temperature range from melting to solidification phase change and they should meet the requirement that the minimum temperature for solidification (activation of crystallization) should be higher than the sink temperature. As explained in this section, in those materials showing a temperature range or supercooling instead of a unique phase change temperature, the temperature range of PCM should be between the heat source temperature and the heat sink temperature. Hopefully, there exist several ways to solve this issue, if it is to be encountered, but none are completely satisfactory. The best solution being always to rely on passive effects, which do not need any further interventions, one uses nucleating agents when they are available. They consist of crystals of other materials that must permit heterogeneous nucleation to occur. Nonetheless, their determination is time consuming and no resilient methodology exists to unambiguously find the pertinent nucleating agent for a given material. The other solution is either to always keep a solidified amount of material, from which crystallization can grow, or to rely on cold fingers, which create local spots where the temperature is so low that it forces the growth of crystal clusters. In both cases, it is clear that a specific system is required, and there's the rub since any system failures can endanger the overall process. Besides, it obliges the designer to also consider control devices and associated algorithms.

The second main issue with PCMs concerns incongruent melting that can exist only for mixtures. It consists of the slow but inevitable separation of the components of a mixture into its basic materials. It happens when the melt is not uniform and leads to at least two solutions with different concentrations. Due to gravity, the crystals will separate and the heavier solution will also "sediment" below the lighter one. Thus, there is a gradual decomposition of the mixture, resulting in the complete separation of all materials that are not mixed anymore. For mixtures that can be subject to incongruent melting, such as salt hydrates, there also exist some solutions. Among these latter, those relying on the addition of more solvent (usually water) must be rejected. Indeed, it leads to a reduction of the storage density. More important, it modifies the phase change interval, which could therefore become very different from the one considered during the design process and/or for the correct operation of the application. The easiest way would be to force the mixing artificially, with the help of any corresponding devices. However, as mentioned before, this imposes the use of a dedicated and specific technology, but it also creates a new risk in case of any dysfunction. The last possibility consists of the use of gelling and thickening agents, which limit the separation of the substances. This solution usually works well, yet there is once again no methodology to find the proper agents. What is more, the phase diagram can be changed and the behavior of the new mixture, containing these agents, must be studied.

Last but not least, the encapsulation of the PCM is a cornerstone of LHS (Farid et al. 2004; Agyenim et al. 2010; Alkilani et al. 2011; Cabeza et al. 2011; Akeiber et al. 2016). This encapsulation can be done in several forms. The most common ones are macro-encapsulation and micro-encapsulation. Besides these solutions, a wide range of different solid stable PCM composites are under development. In all cases, the long-term compatibility of the PCM with the encapsulation, or composite material made of the PCM with another material, need to be ensured. Main compatibility issues come from corrosion phenomena and water/PCM migration through the encapsulation wall. In general, when using inorganic PCMs, corrosion must be studied with metallic encapsulation, meanwhile water migration through plastic encapsulation must be avoided due to the hydrophilic behavior of inorganic salts. Compatibility issues with organic PCMs in general are restricted to the plastic encapsulation. The molecular affinity enables a potential migration of organic molecules through the plastic encapsulation. Adequate plastic composition can be then selected to minimize this phenomenon.

4.3. Commonly used PCMs

The most common commercially available PCMs are mixtures or low grade of purity materials so the phase change takes place over a temperature range instead of a constant temperature. Attending to the chemical nature classification shown in Table 4.1, alkanes and carboxylic acids are the most usual among the organic PCMs. Alkanes are based on n-alkanes of type C_nH_{2n+2} or their mixtures, where the longer the chain ensures higher phase change enthalpy and the melting temperature. Carboxylic acids ($CH_3(CH_2)_{2n}\_COOH$) are included in the subgroup organic non-paraffinic PCMs. They are produced by animals and plants and obtained as hydrolysis to obtain a mixture of carboxylic acids, which has been purified and separated from each other producing fats and oils. They are known as bio-based PCMs. Among the inorganic PCMs, salt hydrates and salts are the most common subgroup in existing PCMs. They are compounds of inorganic salt and water of general formula $AB{\cdot}nH_2O$. The higher latent heat, density and thermal conductivity compared to organic PCMs make salt hydrates the most important group of existing PCMs. However, the potential phase segregation when the material melts due to the density differences of water and inorganic components and the existence of supercooling are the main drawbacks of this group. Metals are included in the inorganic subgroup. Their main features are their higher thermal conductivity and the higher volumetric enthalpy change compared to the PCM presented in this section. Table 4.3 shows the main characteristics of the subgroups presented in this section.

Most common PCMs can be found on the market via chemicals suppliers for high-purity grades or PCM manufacturers or supplier for mixtures, PCM products and composites. Table 4.4 shows examples of the current existing PCM manufacturers and suppliers.

	Organic		**Inorganic**	
	Alkanes	**Carboxylic acids**	**Salt hydrates**	**Metals**
Melting temperature °C	–12°C to 135°C	-7°C to 187°C	–33°C to 120°C	19°C–1,500°C
Density $kg{\cdot}m^{-3}$	750 (l)–900 (s)	878 (l)–1,004 (s)	1,937 (l)–2,180 (s)	5,910 (s)
Phase change enthalpy $kJ{\cdot}kg^{-1}$	60–270	120–250	85–330	25–95
Thermal conductivity $W{\cdot}m^{-1}{\cdot}K^{-1}$	~0.2	~0.2	~0.7	~40
Specific heat (solid, liquid) $J{\cdot}g^{-1}{\cdot}K^{-1}$	2.0 (s) 2.6 (l)	1.7 (s) 2.4 (l)	1.4 (s) 2.3 (l)	1.9 (s) 4.3 (l)

Table 4.3. *The main characteristic according to the PCM classification (based on table 2, p. 135, Khan et al. 2016)*

	Web page	Range of temperature	Observations
Climator	https://www.climator.com/	–21°C to 70°C	Bulk and macro-encapsulated
Cristopia	http://www.cristopia.com/	–64°C to 120°C	Inorganic smaterials (hydrated salts) and organic materials (alkanes and carboxylic acids or bio PCM) Bulk and macro-encapsulated
CRODA	https://www.crodatherm.com/en-gb/products-and-applications/crodatherm-wax/heat-storage-and-recovery	–22°C to 60°C	Bio-based PCM
MikroCaps	www.mikrocaps.com	–21°C to 70°C	Micro-encapsulated
Mikrotek	https://www.microteklabs.com/	–30°C to 57°C	Pure PCM, micro-encapsulated (micronal), slurries and panels
PCM products	http://www.pcmproducts.net/	–100°C to 1,200°C	Organic, inorganic, mixtures Bulk, macro-encapsulated, micro-encapsulated and solid composites
PCP Australia	https://pcpaustralia.com.au	–21°C to 29°C	Bulk and macro-encapsulated
Puretemp	https://www.puretemp.com	–37°C to 150°C	Bulk
Rubitherm	https://www.rubitherm.eu/en/	–50°C to 100°C	Organic, inorganic, mixtures Bulk, macro-encapsulated, micro-encapsulated, emulsions and solid composites

SavEnrg	https://rgees.com/products/pcm-packs	−26°C to 65°C	Macro-encapsulated
Shanghai Tempered Entropy New Energy Co., Ltd	http://www.chinapcm.com/profile	−35°C to 65°C	Organic and inorganic Bulk, micro-encapsulated and emulsions
TEAPPCM	http://www.teappcm.com/products.htm	−50°C to 89°C	Bulk and macro-encapsulated
Tempack	https://www.tempack.com	−4°C to 37°C	Eutectic Macro-encapsulated

Table 4.4. *Example of the current existing PCM manufacturers and suppliers*

4.4. Tecno-economic evaluation

Before tackling this difficult question, it is important to mention that, as far as economic indicators are involved, numerous uncertainties can appear due to the numerous assumptions that must be made. This remark is particularly relevant for thermal energy storages, and even for energy storages in general, since they do not benefit from the same interests and market penetrations that other technologies have up to now.

Consequently, the development of financial models cannot be utterly achieved, especially because the possible boons and associate earnings of latent storages are still under investigation (IRENA 2013; Glatzmaier 2018). However, the main key elements are going to be presented to help interested readers in the development and further improvement of such methods (Rathgeber 2015). To perform such an analysis, it is first needed to list the various elements composing the basic capital expenditures and, then, the ones for the operational expenditures.

In the former case, one has first to consider obviously the cost of the PCM but also the one of the tanks that can often be of the same order (though slightly lower usually). Actually, we must then integrate in the price the auxiliary components system that will permit the heat to exhange during the charge/discharge phases; generally, it is a classic heat exchanger. Naturally, the entire system will need electrical ancillaries and various pipes/valves, as well as pumps and possible instrumentations. These components being strongly dependent on the application case, it is common to assume 3–5% of the investment costs. Eventually, the installation costs are added. Let us remark also that, depending on the case, it could be required to buy or rent land. Last but not least, the initial costs generally involve additional fees for lawyers and consultants. Nevertheless, as they are also present for any project, and not typical of LHSs projects, they are not discussed any further.

When one turns to the operational expenditures, a cornerstone will be the thermal cycling capacity of the PCM. Indeed, it could be needed to replace it, between one to several times, along the lifetime of the project. In fact, it is not uncommon to proceed this way (one can think of PV projects where the inverters are changed every 8–12 years), and this allows accepting some thermal degradation, provided that this decrease is not too fast and too important. One advantage of the thermal energy storages, and so of latent storages, is their ease of use since, basically, only fluid loops are involved and their operations and maintenance and costs are limited. Finally, there would also be other expenditures to consider (insurance, rent, etc.), yet they are common to any other projects, so they are not developed.

Next, when all the above items are known, it is possible to study the rentability of the project. As a general rule, the number of performance indicators are limited, though it is still possible to focus on many others (namely, the internal rate of return, the payback period, etc.). To name but a few, the most often encountered are the capacity and power costs (C_E and C_P), and then the levelized cost of energy (LCOE). The first two performance indicators are written as:

$$C_E = \frac{\text{CAPEX}}{E_{max}}$$

$$C_P = \frac{\text{CAPEX}}{P_{max}}$$

They simply quantify the cost of the system, capital expenditures (CAPEX) in a given currency (€ or $), when compared to the total energy (E), which can be stored (in kWh or MWh), or to the maximal heat flow-rate (P) at which the energy can be withdrawn or injected (in kW or in MW). The present estimations oscillate between 20 and 80 €/MWh.

For the LCOE (€ or $), its aim is basically to estimate the cost of energy on the whole duration of the project, considering capital and operating expenditures (OPEX, € or $) in order to propose a fair basis to compare different technological solutions. Moreover, it uses discounted rates to estimate the present value of the future cash flows obtained during the lifetime of the project. Technically, its computation can be annuitized using the annuity factor (AF) or not, which leads to the following definitions:

$$\text{LCOE}=\frac{\text{CAPEX}+\text{OPEX}\times\text{AF}}{E_{tot}\times AF}$$

$$\text{LCOE}=\frac{\text{CAPEX}+\sum_{i=1}^{LT}\frac{C_i}{(1+r)^i}}{\sum_{i=1}^{LT}\frac{E_i}{(1+r)^i}}$$

where r is the discount rate and the AF for the whole lifetime (LT, years) of the project is given by:

$$\text{AF}=\frac{1-(1+r)^{-LT}}{r}$$

However, in the case of energy storage, this basic definition can underestimate the boons of the energy efficiency arising from the fact that several cycles can be done on a given duration (that is say, all things being equal, especially the discount rate, interest rate, income taxes, etc.). As a result, depending on whether the amount of energy corresponds to the total sum of the energy outflows or entrances (taking then a storage efficiency of 90%) or only to the maximum storage capacity, then the levelized cost of storage can be between \$100 and \$400 per MWh. Added to this, there exist large uncertainties on the PCM costs since, for the same commercial product, one can find various purities, leading to different properties (e.g. amount of storable energy) and very different prices. Besides, these prices can also vary a lot, for instance up to 30%, depending on the bought quantities (kg or tons or several tons).

Finally, let us also mention here that many authors are now more critical with the LCOE since it usually does not consider the final residual value, and more importantly, neither the decommissioning of the installation nor the negative externalities (for instance, the CO_2 cost).

4.5. Emerging alternative materials

The aforementioned materials and their mixtures to be used as PCM already cover a wide range of temperature ranges and storage densities, presenting a variety of chemical properties and compatibilities with the possible encapsulation materials that satisfy a large number of applications. However, the research in materials for thermal energy storage is still very active to foster the expansion of the existing applications or to face new challenges imposed by new applications. The four driven forces to the materials research for thermal energy storage are as follows: (i) to increase thermal conductivity, (ii) to develop form-stable solid composites, (iii) to explore mixtures with melting temperatures and enthalpy changes specific and tunable to certain applications requirements and (iv) to reduce the cost while reducing the environmental impact.

In the previous section, the drawback of low thermal conductivity of PCM was appointed. The main problem is a mismatch of the thermal power requirement when charging or discharging and the heat absorption/release rate from the material. Compared to the other thermal resistances, in the heat transfer fluid or the encapsulation, due to the low thermal conductivity values, the heat transfer by conduction in the LHS materials easily becomes the bottleneck for heat transportation in the system. Therefore, heat exchanger designs are not able to solve the problem or even make the problem worse, and efforts to enhance the heat exchange rate result in systems that are not economically feasible. Material engineering is then required. Attending to the condensed matter physics, phonons are defined as the collective excitation of one elastic structure or arrangement of atoms or molecules. Phonon is designated as a vibration mode and the phonon is the driving force for heat conduction. Low thermal conductivity values are intrinsically present for non-metallic PCMs due to the limited diffusion force of phonons due to the different phonon scatterings inside the material (Wu et al. 2019). Three main approaches are then used to increase the heat transfer rate inside the PCM (Figure 4.7): developing composite materials based on a solid matrix of high conductivity materials and PCM, dispersion of high thermal conductivity particles in the PCM and PCM microencapsulation.

Among organic PCMs studied up to now, sugar alcohols were less developed because most of them exhibit noticeable supercooling and low crystallization rates (Diarce et al. 2015; Palomo 2016). However, they are natural products with higher phase change enthalpies than other organic PCM for a temperature range suitable for many interesting applications such as solar thermal or industrial waste heat storage. These potential applications promoted the research in sugar alcohols mainly focused on their mixtures to find proper eutectic mixtures with melting temperatures at a desired temperature and their crystallization paths and solution to increase the crystallization rates. Besides sugar alcohols, other mixtures as metallic alloys are being explored to fit the TES application at higher temperatures (from 340°C to 1,079°C and melting enthalpy in the range of 155–960 kJ/kg and density values from 2,500 to 8,000 kg/m^3) (El Karim et al. 2019).

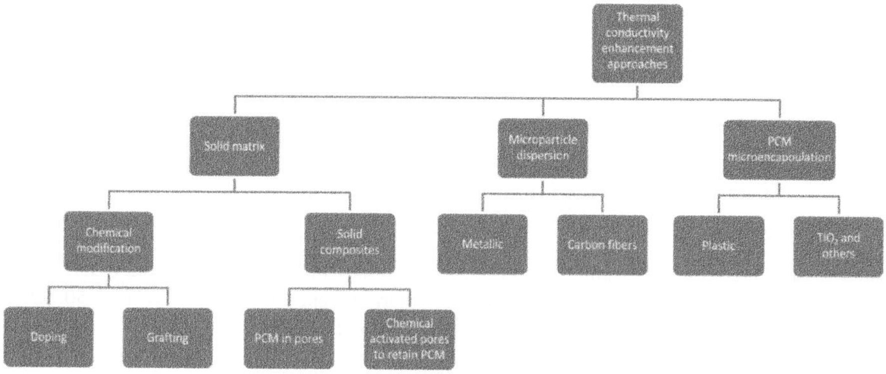

Figure 4.7. *Main approaches found in literature to enhance the heat transfer in PCM*

Non-pure substances normally fall outside the scope imposed by the requirement of a narrow phase change temperature range or high energy storage density. Therefore, for a specific application, only a few materials satisfy the optimal properties and normally the cost of these PCMs is high. Nowadays, the energy market sets energy costs excluding the environmental costs and any other environmental indicators. Thus, the cost of the thermal energy storage system including the material hampers the economic viability for most sustainable energy. Among others, waste and by-products showing a solid–liquid phase change and satisfying the safety requirements are being researched as they can be cheap or even free because industry usually has to bear the costs of their management and disposal.

4.6. References

Agyenim, F., Hewitt, N., Eames, P., Smyth, M. (2010). A review of materials, heat transfer and phase change problem formulation for latent heat thermal energy storage systems (LHTESS). *Renewable and Sustainable Energy Reviews*, 14(2), 615–628.

Akeiber, H., Nejat, P., Majid, M.Z.A., Wahid, M.A., Jomehzadeh, F., Famileh, I.Z., Calautit, J.K., Hughes, B.R., Zaki, S.A. (2016). A review on phase change material (PCM) for sustainable passive cooling in building envelopes. *Renewable and Sustainable Energy Reviews*, 60, 1470–1497.

Alkilani, M., Sopian, K., Alghoul, M., Sohif, M., Ruslan, M. (2011). Review of solar air collectors with thermal storage units. *Renewable and Sustainable Energy Reviews*, 15, 1476–1490.

Cabeza, L., Castell, A., Barreneche, C., de Gracia, A., Fernández, A. (2011). Materials used as PCM in thermal energy storage in buildings: A review. *Renewable and Sustainable Energy Reviews*, 15(3), 1675–1695.

Davis, S.H. (2001). *Theory of Solidification*. Cambridge University Press, Cambridge.

Del Barrio, E.P., Cadoret, R., Daranlot, J., Achchaq, F. (2016). New sugar alcohols mixtures for long-term thermal energy storage applications at temperatures between 70°C and 100°C. *Solar Energy Materials and Solar Cells*, 155, 454–468.

Diarce, G., Gandarias, I., Campos-Celador, A., Garcia-Romero, A., Griesser, U.J. (2015). Eutectic mixtures of sugar alcohols for thermal energy storage in the 50–90°C temperature range. *Solar Energy Materials and Solar Cells*, 134, 215–226.

Dincer, I. and Rosen, M. (2002). *Thermal Energy Storage: Systems and Applications*. John Wiley & Sons, Hoboken, NJ.

El Karim, Y., Grosub, Y., Faikb, A., Lbibba, R. (2019). Investigation of magnesium-copper eutectic alloys with high thermal conductivity as a new PCM for latent heat thermal energy storage at intermediate-high temperature. *Journal of Energy Storage*, 26, 100974.

Farid, M.M., Khudhair, A.M., Razack, S.A.K., Al-Hallaj, S. (2004). A review on phase change energy storage: Materials and applications. *Energy Conversion and Management*, 45(9/10), 1597–1615.

Garg, H.P., Mullick, S.C., Bhargava, A.K. (eds) (1985). Latent heat or phase change thermal energy storage. In *Solar Thermal Energy Storage*. Springer, Berlin. doi: 10.1007/978-94-009-5301-73.

Ge, H., Li, H., Mei, S., Liu, J. (2013). Low melting point liquid metal as a new class of phase change material: An emerging frontier in energy area. *Renewable and Sustainable Energy Reviews*, 21, 331–346.

Gil, A., Medrano, M., Martorell, I., Lázaro, A., Dolado, P., Zalba, B., Cabeza, L.F. (2010). State of the art on high temperature thermal energy storage for power generation. Part 1–Concepts, materials and modellization. *Renewable and Sustainable Energy Reviews*, 14(1), 31–55.

Glatzmaier, G. (2018). Developing a cost model and methodology to estimate capital costs for thermal energy storage. Technical Report NREL/TP-5500-53066. National Renewable Energy Laboratory, Golden, CO.

Glicksman, M.E. (2011). *Principles of Solidification. An Introduction to Modern Casting and Crystal Growth Concepts*. Springer, New York.

Huggins, A. (2010). *Energy Storage*. Springer, Berlin.

IRENA (2013). Thermal energy storage. Technology brief. International Renewable Energy Agency.

Iten, M. and Liu, S. (2014). A work procedure of utilising PCMs as thermal storage systems based on air-TES systems. *Energy Conversion and Management*, 77, 608–627.

Jegadheeswaran, S. and Pohekar, S.D. (2009). Performance enhancement in latent heat ther-mal storage system: A review. *Renewable and Sustainable Energy Reviews*, 13(9), 2225–2244.

Khan, Z., Khan, Z., Ghafoor, A. (2016). A review of performance enhancement of PCM based latent heat storage system within the context of materials, thermal stability and compatibility. *Energy Conversion and Management*, 115, 132–158.

Khudhair, A.M. and Farid, M.M. (2004). A review on energy conservation in building applications with thermal storage by latent heat using phase change materials. *Energy Conversion and Management*, 45(2), 263–275.

Kuravi, S., Trahan, J., Goswami, D.Y., Rahman, M.M., Stefanakos, E.K. (2013). Thermal energy storage technologies and systems for concentrating solar power plants. *Progress in Energy and Combustion Science*, 39(4), 285–319.

Kuznik, F., David, D., Johannes, K., Roux, J-J. (2011). A review on phase change materials integrated in building walls. *Renewable and Sustainable Energy Reviews*, 15, 379–391.

Mehling, H. and Cabeza, L.F. (2008). *Heat and Cold Storage with PCM. An Up-to-date Introduction into Basics and Application*. Springer, Berlin.

Mullin, J.W. (2001). *Crystallization*. Butterworth-Heinemann, Oxford.

Palomo Del Barrio, E., Cadoret, R., Daranlot, J., Achchaq, F. (2016). New sugar alcohols mixtures for long-term thermal energy storage applications at temperatures between 70°C and 100°C. *Solar Energy Materials and Solar Cells*, 155, 454–468. doi: 10.1016/j.solmat. 2016.06.048.

Paksoy, H.Ö. (ed.) (2007). *Thermal Energy Storage for Sustainable Energy Consumption. Fundamentals, Case Studies and Design*. Springer, Berlin.

Rathgeber, C., Lävemann, E., Hauer, A. (2015). Economic top-down evaluation of the costs of energy storages – A simple economic truth in two equations. *Journal of Energy Storage*, 2, 43–46. doi: 10.1016/j.est.2015.06.001.

Sterner, M. and Stadler, I. (2019). *Handbook of Energy Storage. Demand, Technologies, Integration*. Springer-Verlag, Berlin.

Su, W., Darkwa, J., Kokogiannakis, G. (2015). Review of solid-liquid phase change materials and their encapsulation technologies. *Renewable and Sustainable Energy Reviews*, 48, 373–391.

Tatsidjodoung, P., Le Pierrès, N., Luo, L. (2013). A review of potential materials for thermal energy storage in building applications. *Renewable and Sustainable Energy Reviews*, 18, 327–349.

Waqas, A. and Ud Din, Z. (2013). Phase change material (PCM) storage for free cooling of buildings – A review. *Renewable and Sustainable Energy Reviews*, 18, 607–625.

Wu, S., Yan, T., Kuai, Z., Pan, W. (2019). Thermal conductivity enhancement on phase change materials for thermal energy storage: A review. *Energy Storage Materials*, 25, 251–295.

Yuan, Y., Zhang, N., Tao, W., Cao, X., He, Y. (2014). Fatty acids as phase change materials: A review. *Renewable and Sustainable Energy Reviews*, 29, 482–498.

5

Engineering Phase Change Materials to Improve Their Properties and Broaden Applications

Stefania DOPPIU[1] and Elena PALOMO DEL BARRIO[1,2]
[1] *CIC energiGUNE, BRTA, Vitoria-Gasteiz, Spain*
[2] *Ikerbasque, Bilbao, Spain*

5.1. Introduction

Compared to the ideal PCM described in Chapter 3, many of the studied PCMs exhibit one or several undesirable properties. Table 5.1 summarizes frequently encountered problems with examples of concerned material groups. To avoid or to mitigate such problems, several approaches have been proposed. For instance, adding proper nucleating agents usually reduces the amount of PCM subcooling. In addition, uniformly distributing the nucleating agents throughout the phase-change media helps to increase the rate of crystallization in PCMs with slow crystal growth. Moreover, segregation and sedimentation problems appearing in PCMs with incongruent melting could be avoided by adding gelling or thickening agents such as cellulose derivatives, silica gel, sepiolite, diatomaceous earth and cross-linked polymeric hydrogels. However, PCM confinement into micro-/nanoscale frameworks is the approach attracting the greatest interest because of the ability of the resulting novel materials to face most of the problems encountered in solid–liquid PCMs.

Heat and Cold Storage 1,
coordinated by Pierre ODRU, Elena PALOMO DEL BARRIO. © ISTE Ltd 2024.

The development of micro-/nanoconfinement techniques was primarily motivated by the problem of low thermal conductivity of most organic PCMs and that of liquid leakage that affects all solid–liquid PCMs. However, PCMs confinement can also provide many other advantages such as protection from external environment, local management of volume changes, flammability reduction, mitigation of corrosion issues, overcoming phase separation and segregation phenomena, and easier handling and integration into the application.

There are two main strategies to obtain micro-/nanoconfined PCMs, which are usually referred to as micro-/nanoencapsulation and shape-stabilization in the literature. In the first one, the PCM is enclosed in a two-dimensional shell which acts as a container, resulting in nanocapsules (<1,000 nm) or microcapsules (<1 mm) that prevent liquid leakage and protect the PCM against the environment. In the second one, the PCM is infiltrated or intercalated into a micro-/nanoporous material which supports the PCM, avoiding liquid leakage and providing shape and mechanical stability to the material above the melting point of the PCM. Besides these primary effects, the micro-/nanoconfinement can improve the performance of PCMs in several additional ways. For example, heat transfer can be significantly enhanced due to the smaller domains of PCM and larger specific surface area for heat exchange. Moreover, when a conductive shell or supporting material is used, the apparent thermal conductivity of the PCM can be significantly increased. The smaller dimension of the PCM is also efficient in preventing phase separation phenomena characterizing incongruent melting, and larger specific surface area also contributes to promote heterogeneous nucleation, thus mitigating subcooling. The flexibility of the shell or that of the supporting materials, as well as the control of void space within the composite, can accommodate the volume changes of the PCM during melting/crystallization processes. In addition, the surface energy and the surface chemistry of encapsulating/supporting materials, together with small confinement dimensions, can have significant impact on thermodynamics, offering the possibility to shift phase change temperatures and to modulate the latent heat.

The R&D in this area has been increasing tremendously over the last 20 years. A large number of micro-/nanoencapsulated PCMs (ME-/NE-PCMs) and shape-stabilized PCMs (SS-PCMs) have been proposed and studied, covering a wide variety of PCMs, shells, porous materials and characteristic dimensions and geometry of the confinement space (Figure 5.1). A selection of recent review articles on ME-/NE-PCMs and SS-PCMs is provided in Table 5.2. They represent a great contribution to the collection and briefing of research studies on these novel materials, which also provide assistance for further developments and applications. This chapter gives a short overview of the state-of-the-art and trends on this exciting topic.

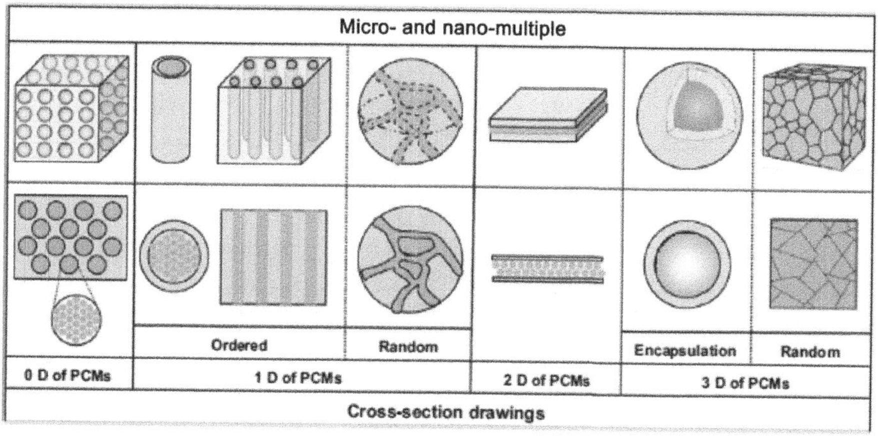

Figure 5.1. *Examples of characteristic dimensions and morphologies of PCMs embedded into porous or shell structures (Zhang and Wang 2018): 0D (zero-dimensional quantum dots or nanoclusters of PCM dispersed into a supporting matrix), 1D (one-dimensional ordered or disordered porous media enclosing a PCM), 2D (two-dimensional layered structures supporting a PCM), 3D (three-dimensional core–shell structures or porous supports enclosing a PCM). For a color version of this figure, see www.iste.co.uk/odru/heat1.zip*

Problem	Consequences	Examples of concerned materials
Low thermal conductivity	Over-sized heat exchangers – Increased investment cost	Non-metallic PCMs (e.g. organic PCMs and most inorganic salts)
Leakage	Limits implementation possibilities and, therefore, applications	All solid–liquid PCMs
High volume change on phase transition (ca. 10% or higher)	Increases the complexity of the storage vessel and that of the heat exchanger, reducing effective energy density and increasing investment cost	Most solid–liquid PCMs, especially inorganic ones
Incongruent melting	Phase separation and sedimentation in containers, causing serious technical problems. Requires specific storage system design, which reduces effective energy density and increases cost	Out-eutectic mixtures such as most inorganic salt hydrates

Poor nucleation ability	Significant subcooling during freezing. If acceptable, it makes difficult the control of the system discharge and reduced recovered heat	Salt hydrates PGE, polyols Polyhydric alcohols
Poor long-term stability	Short life-time. Unacceptable if there is no means of prevention, as for instance using leak-proof containers with inert gas atmospheres	Sugar alcohols Carboxylic acids Urea and urea-based mixtures
Incompatibility with construction materials (corrosion or other)	Reduces the choice of construction materials and/or requires specific protection, thus leading to higher investment cost	Salt hydrates Anhydrous salts Paraffins
Flammability	Limits the applicability, especially in buildings. Requires the incorporation of flame retardant agents to the PCMs	Paraffins In a lesser extent, fatty acids and their derivatives
Non-renewable materials, poor environmental performances	Resources depletion and other negative effects that jeopardize the sustainability	Paraffins Salt hydrates

Table 5.1. *Frequently encountered problems, consequences and examples of concerned PCM groups*

Reference	Topic of the review
Fan and Khodadadi 2011	Thermal conductivity enhancement of PCMs
Kenisarin and Kenisarin 2012	Form-stable PCMs
Fang et al. 2014	Preparation, thermal properties and applications of SS-PCMs
Raam Dheep 2014	Influence of nanoconfinement in PCM properties
Khadiran et al. 2015	Micro-/nanoencapsulated PCMs and SS-PCMs
Liu et al. 2015	Nano-encapsulated PCMs
Liu et al. 2016	Preparation and heat transfer enhancement in ME-PCMs
Lv et al. 2017	Clay mineral-based SS-PCMs
Milián et al. 2017	Encapsulation techniques for inorganic PCMs
Shchukina et al. 2018	Nanoencapsulation of PCMs
Zhang and Wang 2018	Thermodynamics of PCMs in confined spaces
Aftab et al. 2018	Nanoconfined PCMs
Liu et al. 2019b	Design of microencapsulated PCMs
Huang et al. 2019	SS-PCMs based on porous supports
Mitran et al. 2021	SS-PCMs based on porous silica nanomaterials
Hu et al. 2022	Fabrication of organic SS-PCMs

Table 5.2. *Selection of recent reviews on ME-PCMs and SS-PCMs*

5.2. Micro-/nanoencapsulated PCMs

Micro-/nanoencapsulation consists of coating or creating a continuous film around individual particles or droplets of PCM to produce capsules at micrometer to milliliter size (<1,000 μm), known as a microcapsule, or capsules at nanometer size (<1,000 nm), known as nanocapsules. The capsules exhibit a core–shell structure. The PCM forms the core and provides latent heat storage capacity. Depending on the PCM and the deposition process of the shell, the core can be continuous (mononuclear capsules) or divided into many separate regions (polynuclear capsules). The shell is made of a solid polymeric, inorganic or hybrid material and acts as a container to protect the PCM from external environment and to prevent liquid leakage. Micro-/nanocapsules can exist in several shapes: spherical, tubular, oval or irregular. Most common are spherical capsules and fibers (Figure 5.2).

ME-/NE PCMs have great practical interest because they allow expanding the application fields of the PCMs. Indeed, PCM micro-/nanocapsules provide a reliable way for mixing PCMs with polymers and other structural materials, which reduces its flammability, corrosiveness or toxicity while protecting the core material from environmental influence. The application of ME-/NE PCMs in the textile industry is one of the oldest ones (Nelson 2002). The incorporation of ME-/NE PCMs into fabrics and fibers of snowsuits, trousers, ear warmers, boots, gloves, etc., has proven to be efficient in providing extra-protection from extreme cold weather. They are also used in automotive, agriculture and medical textiles.

Another interesting application of ME-/NE PCMs is in the slurry industry. PCM-based slurries with high latent heat are used for cooling and heating applications due to their good performance as a heat transfer fluid and thermal energy storage medium (Tay et al. 2017). In building applications, PCM-based walls or wallboards can be achieved by incorporating ME-/NE-PCMs into construction materials such as gypsum, cement, concrete, wood, etc., to improve the building energy performance (Konuklu et al. 2015). Indeed, the incorporated PCMs increases the thermal mass of the wall so as the indoor temperature swings is reduced and, therefore, indoor thermal comfort is increased while energy consumption is reduced.

Moreover, ME-/NE-PCMs can also be used to improve the thermal performance of insulation materials such as polyurethane foams (Borreguero et al. 2013). As discussed below, with recent progress in the functionalization of ME-/NE-PCMs, new fields of application continue to emerge, such as solar-to-thermal energy storage, electrical-to-thermal energy storage and biomedicine (Arshad et al. 2019).

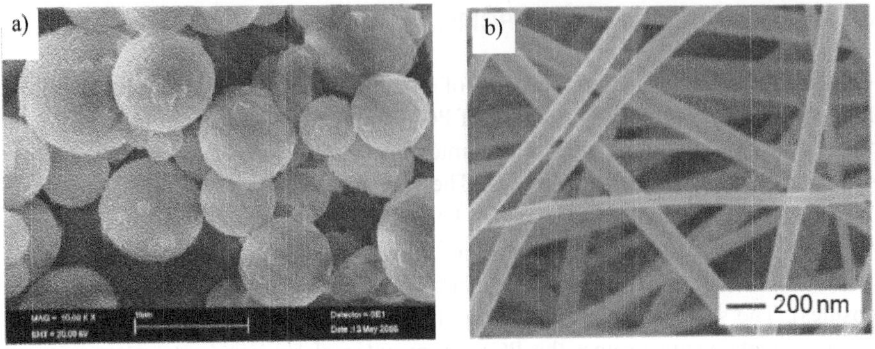

Figure 5.2. *SEM images of spherical-shaped PCM microcapsules (Song et al. 2007) and PCM encapsulated into nanofibers (Mc Cann et al. 2006)*

5.2.1. *Shell materials and encapsulation methods*

The shell material and the method used to create the shell have strong influence on the structural, thermo-physical and mechanical properties of the capsules. The shell materials play a key role in determining the sealing effectiveness and thermal energy storage/release performance of the capsules, because they not only can provide a barrier and protection for the PCM core, but also can supply a huge heat transfer area. An ideal shell material should meet several criteria: (1) chemical compatibility with the PCM core, with no reactions that could engage the chemical stability of the capsule over time; (2) sufficient tightness and mechanical strength to prevent PCM from leakage and support volume changes of the PCM during the melting/crystallization cycles; (3) a good thermal stability to provide high enough upper operation temperature for the encapsulated PCM; (4) a good anti-osmosis property to ensure a long-term working lifespan for the PCM; and (5) a high thermal conductivity to facilitate heat transfer between the PCM core and its environment.

Shell materials can be classified according to their chemical nature into three categories of organic, inorganic and organic–inorganic hybrid materials:

– *Organic shells*: They are usually made up of natural or synthetic polymeric materials and are characterized by good sealing properties and structural flexibility, which offer excellent resistance to the PCM volume changes during phase transitions. Moreover, polymers are cheap, lightweight, easy to process and compatible with many PCMs. Commonly used organic shell materials are melanine-formaldeyde (MF) resin, urea-formaldeyde (UF) resin and acrylic resin. The advantages of MF resin are low cost, good chemical compatibility with most organic PCMs and good thermal stability. Among acrylic resins, the copolymers of metacrylate can be highlighted because they are nontoxic, easy to prepare and have good thermal stability and chemical resistance. Although less common, many other synthetic polymers (e.g. PMMA, PU, PS, and PAN) and bio-based polymers (e.g. gelatin/acacia, gelatin/Arabic gum, ethyl cellulose and methyl cellulose) have also been successfully used as shell materials. We refer interested readers to Liu et al.'s (2019) recent review, which provides a comprehensive map of organic materials employed for PCM encapsulation.

– *Inorganic shells*: Due to the flammability, low thermal conductivity and poor mechanical properties of the polymeric shell materials, inorganic material have been gradually employed for PCM encapsulation in recent years. Compared with organic shells, inorganic shells generally have higher mechanical strength and thermal conductivity. However, they are generally highly rigid and brittle, meaning that the effects of geometry, diameter, shell thickness and void fraction of the resulting capsules are more significant compared to those of polymeric shells and become important technical parameters for design and fabrication of SS-PCMs. In general, smaller diameter, thin-walled spherical capsules provide better structural stability. Silica (SiO_2), zinc oxide (ZnO), titanium dioxide (TiO_2) and calcium carbonate ($CaCO_3$) have been extensively investigated as inorganic shell materials (Aftab et al. 2018; Liu et al. 2019; Peng et al. 2020). Silica is the most commonly used because of its advantages of high thermal conductivity and easy preparation. However, compared to silica shells, $CaCO_3$ shells have higher rigidity and better compactness. The advantages of metal oxide ZnO and TiO_2 shells are their multifunctional properties including catalytic, photochemical and antibacterial properties (Li et al. 2015; Liu et al. 2019b).

– *Organic-inorganic hybrid shells*: This type of shell material emerges to overcome the drawbacks and combine the advantages of organic and inorganic materials (Umair et al. 2019; Liu et al. 2019). Polymeric materials, such as PMMA and PMF doped with SiO_2 or TiO_2, are the most widely used materials in organic inorganic hybrid shells (Palkovits et al. 2005; Li et al. 2019). The inorganic materials serve to enhance the mechanical rigidity, thermal stability and thermal conductivity, while polymers offer structural flexibility. The incorporation of carbon-based materials (e.g. expanded graphite (CG), carbon nanotubes (CNTs) and graphene oxide (GO)) into polymeric shells has also attracted much attention over the last decade because of the higher thermal conductivity of carbon-based materials compared to oxides (see Liu et al. (2019a) and references therein). Moreover, carbon-based hybrid shells provide the capsules with additional abilities such as photothermal conversion.

Microcapsules can be obtained by physical, chemical and physical–chemical methods (Figure 5.3). *Physical methods* only involve physical processes such as drying, dehydration and adhesion. Low-cost, versatility and scalability are the main assets of the spray drying method, which produces microcapsules of 5–5,000 μm in size. Solvent evaporation (0.5–1,000 μm capsules) is also a low-cost process but the use is restricted to lab-scale production because of low yield. *Chemical methods* employ polymerization or a condensation process of monomers, oligomers or prepolymers as raw materials to form shells at the oil–water interface. The differences between in situ, interfacial, suspension and emulsion polymerization are shown in Figure 5.3. The capsules range from 0.5 to 1,200 μm. *Physical–chemical methods* combine physical processes (e.g. phase separation, heating and cooling) with chemical processes (e.g. hydrolysis, cross-linking and condensation) to achieve microencapsulation. Coacervation (2- to 1,200-μm capsules) and sol-gel (0.2- to 20-μm capsules) encapsulation are the most popular physical–chemical methods. The latter is usually used to synthetize inorganic shells and it is an economical and mild process. To get a detailed description and deeper analysis of the different microencapsulation techniques, we recommend the recent review articles by Kadhiran et al. 2015; Milián et al. 2017; and Liu et al. 2019b.

Compared to microencapsulation, there is relatively less work done in nanoencapsulation. However, with the spectacular progress of nanotechnologies in the last decades, the interest in nanoencapsulation is growing (Aftab et al. 2018; Shchukina et al. 2018; Liu et al. 2019b). Much better heat transfer is the primary benefit expected from nanoencapsulation. Indeed, smaller capsules greatly increase the surface-area-to-volume-ratio, which improves heat transfer. PCM-based core–shell nanostructures with organic shells can be obtained by polymerization methods, while inorganic coating is achieved by sol-gel method. The miniemulsion polymerization is the most adopted polymerization technique. Apart from spherical core–shell nanostructures, PCM encapsulation into polymer nanofibers has attracted much interest during last decade. Electrospinning is the most commonly used method to produce PCM nanofibers because of the unique properties of resulting nanofibers such as light weight, small diameter, controllable porous structure, high-surface-to-volume ratio, flexibility and superior mechanical properties (e.g. stiffness and tensile strength). The fundamentals of electrospinning can be found in pioneering work by Doshi and Renerker (Doshi and Renerker 1995). PCM-blended polymer nanofibers are obtained by single-phase electrospinning method, which involves spinning of a mixture containing PCM and supporting polymer, while core–shell PCM nanofibers are obtained by the coaxial electrospinning method.

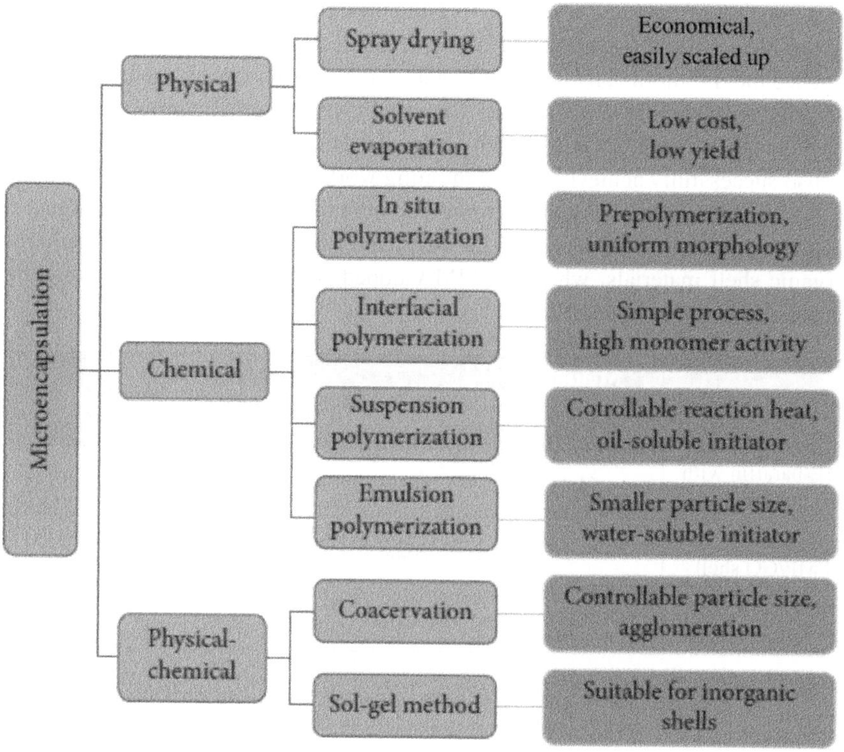

Figure 5.3. *Classification of microencapsulation methods (Peng et al. 2020)*

5.2.2. *Micro-/nanoencapsulated PCMs*

As reported in recent review papers (Khadiran et al. 2015; Liu et al. 2015, 2016, 2019b; Milián et al. 2017; Aftab et al. 2018; Shchukina et al. 2018), hundreds of articles have been published on ME-/NE-PCMs since 1990. They reveal that organic PCMs have attracted more attention than inorganic ones, probably because their encapsulation is easier.

5.2.2.1. *Organic PCMs*

Among organic PCMs, n-alkanes and paraffin waxes are the most commonly used core PCMs, although fatty acids, fatty alcohols and PEG have also been successfully encapsulated. Polymers are the most frequently used materials for organic PCMs encapsulation. It has been shown that the type of polymeric shell plays a key role in determining the thermal properties of the capsules, whereas the characteristics of the shell such as molecular weight and distribution, crosslinking degree, thickness and tightness, stiffness, mechanical strength and thermal dilatation influence the thermo-mechanic stability and durability. Moreover, the microcapsules with polymeric shell in small size are found to have a better structural stability. The encapsulation of organic PCMs with inorganic or organic–inorganic hybrid shells has also successfully achieved. The compiling work by Liu et al. (2019b) provides a list of more than 70 micro-/nanoencapsulated organic PCMs with either inorganic or organic–inorganic hybrid shell. SiO_2, $CaCO_3$ and TiO_2 are the most commonly used inorganic shell materials, whereas PMMA doped with oxides is the most frequent organic–inorganic hybrid shell.

ME-/NE-PCM	Melting temperature (°C)	Latent heat of melting (J/g)	PCM loading (wt%)	Reference
ME-paraffin with PUF shell	54.4	201.2	97.9	(Ji 2014)
ME-paraffin with PMF/GO shell	44.2	202.8	93.9	(Zhang et al. 2017)
ME-stearic acid with SiO_2 shell	53.3	171	90.6	(Chen et al. 2013)

Table 5.3. *Examples of core–shell encapsulated organic PCMs with exceptionally high PCM loading*

The proportion between PCM core and shell material is also an important parameter to determine the energy performance and structural stability of the capsules. Relatively high weight percentage of PCM core leads to higher heat storage capacity but poor structural stability and vice versa. Thus, the mass loading of PCM needs to be optimized with respect to the mechanical strength of coating materials. The PCM loading in micro-/nanocapsules generally ranges from 25 to 80 wt%, meaning a loss of storage capacity of 75–20% compared to bulk PCM. PCM loading above 80 wt% is rarely achieved, although several authors have reported exceptionally high values (Table 5.3).

As already mentioned, nanoencapsulation provides increased surface-area-to-volume-ratio compared to microencapsulation, thus showing better heat transfer performance. Structural stability is also generally enhanced when reducing the size of the capsules. Moreover, the flowability of nanocapsules is better than that of microcapsules, which facilitate their integration in construction materials, slurries, etc. However, nanoencapsulation often tends to amplify subcooling phenomena. As can be seen in Table 5.4., nano-encapsulated paraffin waxes display significant degree of subcooling (approximately 7°C), while it is negligible in bulk paraffin waxes.

PCM	Shell	Encapsulation method	PCM ratio (%)	T_m (°C)	T_c (°C)	ΔH (J/g)
Paraffin wax	PS	Miniemulsion polymerization	50	27.1	19.6	110.1
Paraffin wax	PU	Suspension polymerization	51.5	55.3	48.0	102.2
Paraffin wax	SiO2	Sol-gel	47.4	27.1	19.6	110.1
Paraffin wax	PMMA- SiO_2	Interfacial polymerization	57.4	26.8	19.8	69.9

Table 5.4. *Examples of the effect of paraffin wax's nano-encapsulation on the subcooling degree (source: Aftab et al. 2018). Tm: melting temperature; Tc: crystallization temperature; ΔH: enthalpy of melting*

Despite the progress made in the micro/nanoencapsulation of organic PCMs, the products available on the market are still scarce. We can mention the microcapsules of paraffin waxes with polymeric shells offered by BASF (Micronal PCM, DS 5000 – DS 5040), Rubitherm GmbH (RT 18 – RT35) and Microteck Laboratories, Inc. (MPCM 22P-29P), with melting point around room temperature and latent heat values ranging from 110 to 195 J/g.

5.2.2.2. *Inorganic PCMs*

Contrary to organic PCMs, whose encapsulation is well established by different encapsulation techniques, there is much less experience in the encapsulation of inorganic PCMs. Among inorganic PCMs, salt hydrates with melting point below 100°C are the most frequently studied core materials. Salt hydrates are difficult to encapsulate due to their hydrophilicity, tendency to alter water content and surface polarity. Encapsulation of inorganic PCMs with polymeric shells is expensive and limited to only inverse emulsion and addition polymerization reactions (Hassan et al. 2016; Milián et al. 2017; Shchukina et al. 2018). On the contrary, salts hydrates are properly encapsulated by using either organic–inorganic hybrid polymers (ORMOCER®s – Impermeable to water, biodegradable inorganic–organic hybrid materials developed by Fraunhofer ISC, Germany) or inorganic materials (SiO_2, TiO_2) as shell materials (Shchukina et al. 2018). In the former case, a co-extrusion technique combined with UV-curing is used to encapsulate the PCM, whereas physical–chemical sol-gel method is employed in the latter case.

PCM	Shell	Encapsulation method	PCM ratio (%)	T_m (°C)	T_c (°C)	ΔH (J/g)
$Na_2SO_4 \cdot 10H_2O$	SiO_2	Sol-gel emulsion	58	37	−15	180.7
$Na_2SO_4 \cdot 10H_2O$	PECA	Inverse emulsion polymerization	61	32	17	138.6
$Mg(NO_3)_2 \cdot 6H_2O$	PECA	Inverse emulsion polymerization	54	92	82	88.4
$CaCl_2 \cdot 6H_2O$	VOOH	Template method	–	30	–	100

Table 5.5. *Examples of core–shell nano-encapsulated salt hydrates (source: Aftab et al. 2018). T_m: melting temperature; T_c: crystallization temperature; ΔH: enthalpy of melting*

Several examples of successfully encapsulated salt hydrates are provided in Table 5.5. Besides usual advantages of encapsulation such as antileakage effect, PCM protection and enhanced heat transfer, encapsulated salt hydrates display a major advantage compared to the pure PCM. Indeed, it is observed that micro-/nanoencapsulation overcomes segregation and phase separation issues of slat hydrates, providing excellent cyclability. However, as shown in Table 5.5, subcooling remains a concern.

To date, examples of encapsulation of other types of inorganic PCMs are scarce, limited to the encapsulation of some metal-based PMs (Nomura et al. 2015; Nomura 2017) and high melting temperature anhydrous salt $NaNO_3$ (Li et al. 2017).

5.2.3. *Innovative designs for versatile applications*

Multifunctional ME-/NE PCMs are a new class of materials that are capable of performing other functions in addition to storing thermal energy. The original idea was introduced by Fei et al. (2008), who achieved the encapsulation of *n*-octadecane/titania aerosol with anatase TiO_2 shell by a hydrothermal method. The resulting capsules were shown to have photocatalytic and UV barrier activity in addition to thermal energy storage.

This work, which went unnoticed at the time, gave rise to a new line of research related to the encapsulation of PCMs that has been quite active since 2015. A comprehensive review of the work done to date can be found in Aftab et al. (2018), which we briefly summarize here.

PCM capsules provided with additional functions are achieved by design of single, hybrid or multi-layered shells incorporating functional materials such as oxides, metals or semiconductors. Observed additional functions/properties are photocatalytic activity, UV protection, antibacterial or sterilization effects, photoluminescence, magnetic properties, solar-to-thermal conversion, electricity-to-thermal conversion and/or electrical/thermal dual energy storage ability (Figure 5.4). As discussed below, these additional functions extend the application field of PCMs.

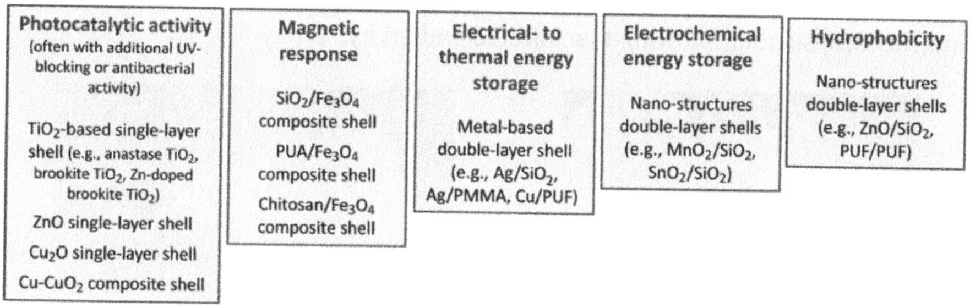

Figure 5.4. *Examples of additional functions of ME-/NE-PCMS that can be achieved by innovative shell design*

The known photocatalytic properties of TiO_2 have made it possible to generate PCM capsules with TiO_2-based shells that show photocatalytic activity and can be used, combined with textiles or construction materials, as thermal regulators with the added advantage of their catalytic power against gaseous pollutants such as NOx. Some of them also have antibacterial properties. Cu_2O-based shells exhibit photocatalytic effects as well, besides high efficiency in photothermal conversion (high solar radiation absorption) and gas sensitivity to some harmful organic gases. PCM core was made of either *n*-alkanes or paraffin waxes in all the cases.

Microencapsulated PCMs with magnetic properties have attracted great interest in both industry and academia due to their potential application for stealth aircraft and other electromagnetic-shielding for military systems. Due to the known magnetic properties of Fe_3O_4, it has been used to produce magnetic ME-PCMs. Fe_3O_4 particles are very sensitive to oxidation and tend to aggregate, making it difficult to fabricate Fe_3O_4 shells.

On the contrary, it has been possible to fabricate hybrid shells in which the Fe_3O_4 particles are integrated into an inorganic (SiO_2) or polymeric (PUF, chitosan) support. Again, *n*-alkanes and paraffin waxes were used as PCM.

Silver-based shells were also investigated due to their high thermal and electrical conductivity properties as well as excellent sterilization effectiveness. As for magnetic microcapsules, ME-PCMs with silver shell could not be obtained from direct formation of a silver shell around the PCM because Ag is so soft that it does not provide enough mechanical stability and rigidity to the capsules. Instead, double-layer shell structures have been proposed, with an inner layer made of SiO_2 or PMMA and a silver outer layer. Paraffin waxes and n-alkanes were used as PCMs. In a similar way, Cu/PUF double-layered shell has been fabricated for paraffin encapsulation showing low infrared emissivity.

a) b)

Figure 5.5. *Microencapsulated PCMs with micro-nanostructured shells: (a) nanoflake-like MnO_2/SiO_2 double-layered shell (Xu et al. 2018); (b) nanoflower-like ZnO/SiO_2 double-layered shell (Sun et al. 2019)*

Micro-/nanostructured shells for PCM encapsulation is another hot topic. PCM microcapsules with nanoflake-like MnO_2/SiO_2 double-layered shell and *n*-docosane as core have been produced by Xu et al. (2018). They prove to have higher capacitance than traditional MnO_2 electrodes while providing efficient thermal management. Figure 5.5(a) shows SEM images of these capsules, where nano-structured features of the shell can be appreciated. Microcapsules of oleic acid-PEG eutectic PCM with mild flake-like SnO_2/SiO_2 double-layered shell

(Imran Hussain et al. 2018) also demonstrated good thermal energy storage and electrochemical performance. Nano-structuring the outer surface of the capsules is also an effective method of achieving superhydrophobic properties. PCM microcapsules with ZnO/SiO_2 double-layer shell produced by Sun et al. (2019) are an example (Figure 5.5(b)). Super-hydrophobicity is achieved by building an outer layer of ZnO in the form of a nanoflower over the silica-based inner layer.

5.3. Shape-stabilized PCMs

SS-PCMs, also referred as to form-stable PCMs, are a new class of materials that emerge as a simpler and cheaper alternative to PCM encapsulation into core–shell structures. Often, they are also more efficient than ME-/NE-PCMs in increasing the apparent thermal conductivity of the PCMs. In SS-PCMs, the PCM is incorporated into a porous support that allows preventing liquid leakage while keeping the shape of the PCM when the temperature is over the melting point. The anti-leakage effect results from capillarity, surface tension, Van der Waals forces and/or hydrogen bonding depending on the size of the pores and chemistry of the PCM and that of the surface of the solid support. In addition to anti-leakage ability, SS-PCMs can exhibit many other advantages compared to bulk PCMs (Table 5.6) such as higher thermal conductivity, local management of the PCM volume changes, mitigation of segregation or phase separation phenomena in PCMs with incongruent melting, subcooling reduction, PCM protection against environmental influence, mitigation of corrosion or flammability issues, etc. Moreover, playing with the size and morphology of the porous space, thermodynamics of the PCM can experiment significant changes, such as shifts of transition temperatures and/or latent heat enhancement, providing a way to tailor their storage performances.

As it is the case for ME-/ME-PCMs, SS-PCMs allow extending the application fields of PCMs as well. For instance, they can be easily incorporated into different construction materials (e.g. gypsum, cement and concrete) for passive heating and thermal comfort in buildings. Moreover, some of them have solar-to-thermal and electricity-to-thermal conversion properties that make them suitable for a new generation of versatile storage systems, whose charge can be achieved by either solar radiation directly impinging on the material or renewable electricity. Because of their anti-leakage ability and thermal energy storage performance, they are also excellent candidates for batteries or microelectronic thermal management or as thermal protection barriers. Most importantly, they offer the possibility of strongly reducing the capital cost of latent heat storage systems, which is currently the most important limitation for the penetration of this technology into the market. Indeed, provided they are compatible with the used heat transfer fluid, SS-PCMs duly shaped and sized could be used as filler material in packed-bed storage systems. In such a way, the need for internal heat exchangers that represents more than 70% of the CAPEX (often more than 85%) is eliminated, thus achieving strong cost reduction and, therefore, competitive and compact solutions against sensible heat storage systems.

Problem	Benefits that can be obtained by using SS-PCMs
Low thermal conductivity	SS-PCMs, especially those with supporting carbon-based material, lead to significant increase in the apparent thermal conductivity of low conductive PCMs.
High volume change during phase transition	Most SS-PCMs have sufficient void space (the porosity is not fully filled by the PCM) to manage the volume changes at a local scale (pore scale), thus simplifying the storage system design.
Incongruent melting	SS-PCMs can prevent the solid phase precipitating out of the liquid during the non-congruent melting, thereby separating macroscopically the phases. Indeed, phase separation is restricted to the microscopic distances determined by the size of the pores in the supporting porous materials.
Poor nucleation ability	Sometimes, the supercooling can be reduced. This occurs when the surface of the porous material provides suitable sites for heterogeneous nucleation. In addition, the higher the specific surface area, the greater the effect on the undercooling.
Poor long-term stability	SS-PCMs protect the PCM against the surrounding environment and, therefore, can contribute to extend its lifetime.
Incompatibility with construction materials (corrosion or other)	As the PCM in porous media is protected from the external environment, the latter is protected from PCMs. Indeed, the contact between the PCM and the construction materials is significantly reduced in SS-PCMs.
Flammability	The incorporation of flame retardant agents can be easily achieved in SS-PCMs. Moreover, many inorganic supporting materials have flame retardant properties.

Table 5.6. *Additional benefits that can be obtained from SS-PCMs compared to bulk PCMs*

The idea of SS-PCMs came after a quite long research period where the first objective was to increase the apparent thermal conductivity of PCMs. To this end, the incorporation of conductive fillers into the PCM was extensively investigated. Most of the published papers deal with the dispersion of metal nanoparticles, metal oxide nanoparticles or carbon-based nanostructures (e.g. nanotubes, nanofibers and nanoplates) into the PCM (Khodadadi et al. 2013). Unfortunately, the resulting composite PCMs often had poor stability because of the difference in density between the fillers and the PCM. Moreover, the thermal conductivity improvement was limited by contact thermal resistances between fillers. Using conductive porous structures, such as metallic foams or graphitized carbon foams, hosts for the PCM were then proposed to overcome such drawbacks (Huang et al. 2019). They demonstrated they were highly efficient in increasing PCMs' apparent thermal conductivity while keeping high energy storage capacity. However, both metal and carbon foams are characterized by rather large pores (millimetres in size), which makes it impossible to prevent liquid leakage. In order to stabilize the shape and avoid leakage, it was therefore necessary to move toward porous media with pore sizes small enough to promote a capillary effect.

In the following sections, we first introduce the different porous supports that have been studied to produce SS-PCMs, as well as the production methods. Then, we focus on the description of most extensively investigated classes of SS-PCMs, namely, porous carbon-based SS-PCMs and porous oxide-based SS-PCMs. Lastly, we offer a comparative analysis of them, as well as a short description of current research trends and market status.

5.3.1. *Overview of porous supports and preparation methods*

A large variety of porous supports have been proposed and studied for SS-PCM production. They differ either in surface chemistry or porous space characteristics or both. The encapsulation efficiency strongly depends on pore size and total porosity of the supporting material as well as on the chemistry of the surface of the porous skeleton and that of the PCM. The volume of pores (total porosity) determines the overall uptake of PCM and thus the energy storage capacity. The role of pore size is critical in determining the possibility of filling the porous space with a PCM. For instance, if the pore aperture is smaller than the overall molecular size of the PCM, the infiltration efficiency would be drastically limited. The pore size also determines the easy crystallization of the PCM inside the host. Indeed, the degree of freedom of movement of PCM molecules within the smaller pores is limited and may prevent the growth of the crystalline phase. It is known that organic PCMs do not crystallize in pores below the size of approximately 5 nm. In this regard, the molecular length of the PCM is also important since smaller molecules would undergo crystallization more easily than larger molecules. In comparison to smaller pores, the larger pores provide enough space for crystal growth due to lower interfacial interactions between the PCM molecules and the host. However, too large pores (approximately

>50 nm) would limit capillary effects so as the anti-leakage ability could significantly drop. Therefore, the most suitable pore size range for SS-PCM production seems to be in the range of approximately 5–50 nm. Regarding subcooling, it is generally reduced in smaller pores because of strong interfacial contact between PCM and host materials and larger surface-to-volume ratios, which will favor heterogeneous nucleation, whereas large pores are usually inefficient in reducing the subcooling. It is also known that nanoconfinement could strongly modify the thermodynamics of the PCM, leading to significant shifts in the transition temperatures, changes of latent heat values and even appearance of new solid phases (Alba-Simionesco et al. 2006). These changes are rarely observed in larger pores, whereas they are common in the smaller ones, where surface-to-volume ratios are high and favor PCM/host interactions. In this regard, the morphology of the pores could also be important, because it can lead to the formation of more or less dense packing of the solid phases depending on the shape and size of the pore as well as that of the PCM molecules. Also, the chemistry of the surface of the pores and that of the PCM plays a crucial role in regard to the direction and significance of the thermodynamic changes (e.g. increasing/decreasing transition temperatures or latent heat values). Generally, a strong attractive interaction between liquid PCM and porous support will lead to increased phase transition temperature, whereas a weak attractive interaction will result in a shift toward lower temperatures. At last, the chemistry of the PCM and that of the surface of the pores could allow the establishing of chemical bonds between them (e.g. hydrogen bridges) that will also contribute to the modification of the PCM thermodynamic properties compared with its bulk counterpart.

Disregarding possible modifications of the thermodynamics of PCM, which are still insufficiently understood, some general criteria can be established to select or design a porous support for its use in the production of SS-PCMs. They are as follows: (1) chemical compatibility with the PCM to warrant durability; (2) good thermochemical stability in the temperature range of use; (3) high volume of pores to maximize PCM loading and, therefore, high energy storage capacity; (4) pore size neither too small nor too big to allow crystallization while preventing leakage; (5) large specific surface area to enhance heat transfer and, eventually, contributing to subcooling mitigation; (6) high thermal conductivity to enhance heat storage/delivery rates; (7) sufficient mechanical strength to support the volume changes of PCM during phase transition processes; and (8) abundant, cheap, safe, environmentally friendly, etc.

According to the chemical nature of the support (Figure 5.6), extensively investigated SS-PCMs can be classified into two main categories: (i) SS-PCMs with an organic supporting material and (ii) SS-PCMs with an inorganic supporting material.

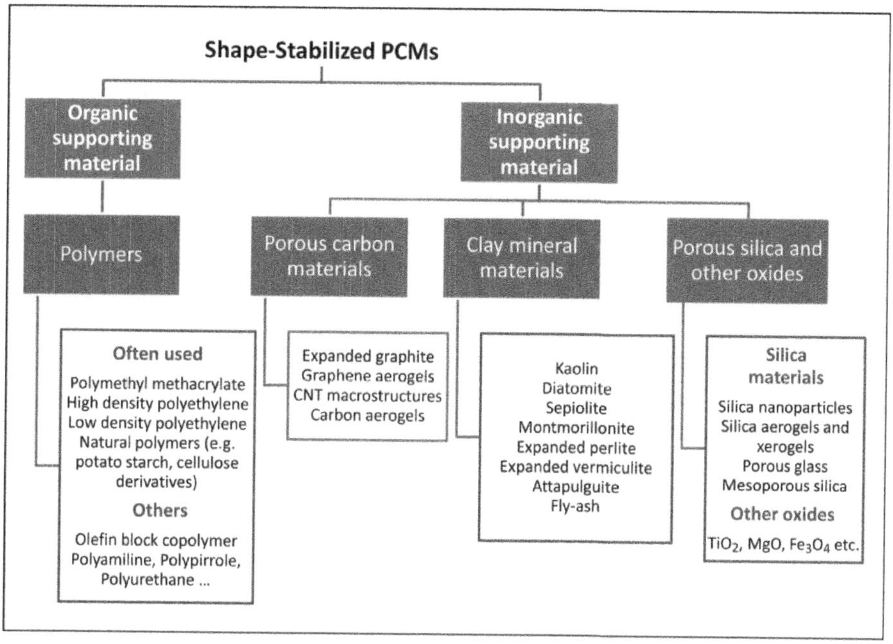

Figure 5.6. *Classification of extensively investigated SS-PCMs according to the nature of the supporting material*

SS-PCMs with an organic support material were the first to be produced and investigated (Huang et al. 2019). These SS-PCMs are obtained from a liquid mixture of PCM and a polymeric material, which when cooled below the glass transition temperature of the polymer solidifies and encapsulates the PCM within a porous polymeric structure. As shown in Figure 5.6, many different types of polymers have been used as support material for organic PCMs (e.g. paraffin waxes, n-alkanes and fatty acids).

Nowadays, the focus has shifted to *SS-PCMs with inorganic support material* (Lv et al. 2017; Aftab et al. 2018; Huang et al. 2019; Mitran et al. 2021). There are several reasons for this. First, there is a wide diversity of porous inorganic materials that offer a broad spectrum of possibilities to optimize or tailor the properties of SS-PCM by simply selecting those with proper chemical-physical and structural/ geometrical characteristics. Moreover, for many of them, it is possible to tailor the porosity (e.g. pore size, volume of pores, connectivity and morphology) as well as to modify the chemistry of the pores' surface to enhance/amplify sought characteristics for the SS-PCM. In general, they are also more conductive than polymers and not flammable. Besides, they allow encapsulating not only organic PCMs, but also salt hydrates and high-temperature molten salts. In addition, among extensively studied inorganic porous supports, there are many which are abundant and cheap. As can be seen in Figure 5.6, inorganic porous supports can be subdivided into two main classes: porous carbon-based SS-PCMs and porous oxide-based SS-PCMs that will be discussed in the next three sections.

The incorporation of the PCM into the inorganic porous structure can be done in several ways. Most of the SS-PCMs are produced by vacuum-assisted infiltration of liquid PCM (molten PCM, PCM solution) into the porous structure of the supporting material. Although less frequent, the solid PCM loading method has also been used. In this case, PCM powders and supporting material are mixed and pressed to form a pellet, which is then submitted to thermal treatment. SS-PCMs using SiO_2- or TiO_2-based frameworks can also be obtained by sol-gel method.

5.3.2. *Porous carbon-based SS-PCMs*

Carbon materials such as EG, CNTs and GO are being extensively investigated as supporting materials for SS-PCMs production (Aftab et al. 2018; Huang et al. 2019). They offer the possibility of PCM confinement in one-dimensional (1D; e.g. PCM-filled CNTs), two-dimensional (2D; e.g. PCM molecules intercalated between GO layers) and three-dimensional (3D; e.g. PCM-filled EG or GO aerogels) frameworks. 1D and 2D confinement are also referred to as tubular confinement and interfacial confinement, respectively, while 3D confinement is also referred to as porous confinement. Comparing interfacial confinement with porous confinement, the former deals with comparatively strong hydrogen bonding and grafting of PCM molecular chains onto GO sheets, whereas the latter keeps liquid PCM packed inside the pores mainly by capillary action.

Tubular and interfacial confinements are yet in an early stage of investigation, with no conclusive results so far on the proven modification of the thermodynamics of the PCMs (Aftab 2010). On the contrary, carbon-based 3D porous materials including porous CNT macrostructures, graphene aerogel, carbon aerogel and EG have been successfully used as supporting material to hold PCMs. The widespread utilization of these materials relates to some outstanding properties. Compared to metal foams, these materials are cheaper, lightweight, stable and compatible with all kind of PCMs. They also have high internal nanoporosity and large specific surface area, contrary to metal foams which are macroporous materials with poor ability to prevent liquid leakage. In comparison with ceramics or clay mineral supports, porous carbon materials have higher thermal conductivity, which is advantageous in accelerating charging/discharging processes. They also have an electrical conductivity suitable for the conversion of electricity into thermal energy, which would make them useful with respect to the objective of electrification of the heat sector for its decarbonization. Moreover, the black color of carbon materials allows sunlight absorption with good photo-thermal conversion efficiency, which make carbon-based SS-PCMs promising materials for solar-thermal conversion and storage. Last but not least, carbon-based 3D porous materials can be easily functionalized and often customized.

5.3.2.1. *Graphene aerogel-based SS-PCMs*

Graphene is a 2D material composed by sp^2-bonded carbon atoms with a hexagonal honeycomb lattice structure. Self-assembly of GO sheets into 3D structures through a reductive hydrothermal process or directly grown by CVD method offers a large number of interconnected open pores that can be filled with liquid PCMs to produce conductive SS-PCMs (Aftab et al. 2018).

Graphene aerogel prepared by conventional hydrothermal method consists of randomly interconnected network of graphene sheets (Figure 5.7(A)), resulting in an isotropic material. Zhong et al. (2013) were the first to produce an SS-PCM by infiltrating isotropic graphene aerosol with octadecanoic acid. The thermal conductivity of the SS-PCM increased up to 2.6 W/m/K with 20 wt% loading of graphene aerogel, which is 14 higher than that of pure stearic acid. The composite also exhibited a high latent heat value of 182 J/g. Equally good results were obtained by Zhang et al. (2016) when incorporating paraffin wax instead of stearic acid to graphene aerogel (Figures 5.7 (A) and (B)). Graphene aerogel can be provided with anisotropic properties when prepared by employing gaseous HCl to in situ solidify ordered GO liquid crystals followed by reduction, supercritical drying and thermal annealing in an inert atmosphere (Li et al. 2016). This results in a GO layered structure (Figures 5.7 (C) and (D)) with superior heat transfer ability compared to isotropic graphene aerosol, thus leading to more efficient SS-PCMs.

To further improve graphene aerogel-based SS-PCMs, Yang et al. (2016) fabricated high-density hybrid graphene aerogel by reducing and self-assembling a mixture of GO and graphene nanoparticles followed by open air-drying. SS-PCM obtained by incorporation of octadecanol with 12 wt% graphene shown thermal conductivity 26 times higher (5.9 W/m/K) than that of pure octadecanol while maintaining high latent heat of melting (203 J/g). Moreover, high-density hybrid graphene aerogel provides SS-PCM with higher mechanical strength. Following the work of Yang et al., other hybrid graphene aerogels have been proposed whose description can be found in Aftab et al. (2018) and Huang et al. (2019).

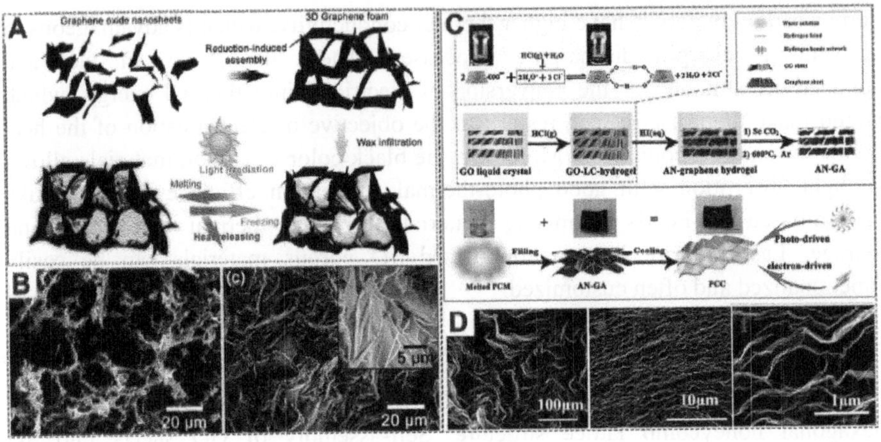

Figure 5.7. *3D porous graphene structures and SS-PCMs. Scheme for the preparation of the 3D porous graphene and wax composite (A). SEM images of porous graphene (left) and the graphene–wax SS-PCM (right) (B) (Zhong et al. 2013) Synthetic illustration of anisotropic graphene aerogel and SS-PCM (C). SEM images of anisotropic graphene aerogel (D) (Li et al. 2016). For a color version of this figure, see www.iste.co.uk/odru/heat1.zip*

5.3.2.2. *SS-PCMs based on porous CNT macrostructure*

CNTs have been extensively investigated as conductive fillers for PCMs' heat transfer enhancement (Khodadadi et al. 2013; Huang et al. 2019). Indeed, single-walled and multi-walled CNTs have extremely high thermal conductivity in their axis direction (SWCNT: 6,600 W/m/K @RT; MWCNT: 3,000 W/m/K @RT). However, their efficiency in increasing the apparent thermal conductivity of PCMs is unsatisfactory due to thermal contact resistances between CNTs dispersed into the PCM. To overcome this limitation, 3D networks of CNTs with higher continuity of the carbon phase would be a solution.

3D carbon nanotube macrostructures include carbon nanotube sponge (CNTS) and carbon nanotube array (CNTA). CNTS is an interconnected network of CNTs prepared by the CVD method (Figure 5.8(B)). It is a flexible, lightweight and highly porous (>99% porosity) structure, which makes it very interesting as support for PCMs (Gui et al. 2010). Using a simple solution infiltration method (Figure 5.8(A)), Chen et al. (2012) obtained a paraffin-CNTS composite with more than 90 wt% loading of paraffin that demonstrated excellent anti-leakage ability and significant thermal conductivity enhancement (6.7 times that of paraffin). In addition, a considerable increase in latent heat was also observed due to the strong π-π interactions between paraffin molecules and the CNT wall.

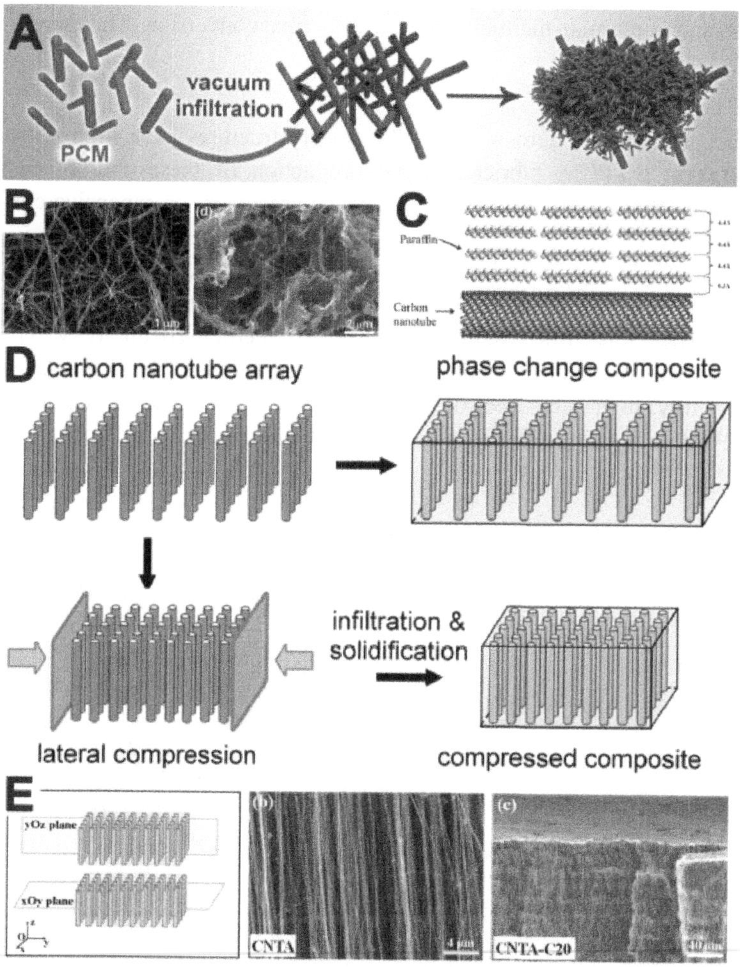

Figure 5.8. *Confining PCM within CNTS (A). SEM images of empty (right) and infiltrated sponge (left) (B) (Chen et al. 2012). Arrangement of the paraffin chains along the nanotube (C). Melt impregnation of C20 into CNTA, or compression of CNTA followed by C20 infiltration (D). SEM images of CNTA (E) (Liu et al. 2013). For a color version of this figure, see www.iste.co.uk/odru/heat1.zip*

CNTA consists of a set of CNTs arranged in parallel (Figure 5.8(D)). Liu et al. (2013) prepared an SS-PCM by filling CNTA with eicosane

(Figure 5.8(E)). They show that the thermal properties of the SS-PCM can be easily modulated by lateral compression of the CNTA (Figure 5.8(D)). Resulting anisotropic SS-PCMs displayed high thermal conductivity in the CNTs' axis direction and good solar-thermal and electro-thermal conversion efficiency.

5.3.2.3. *Carbon aerogels*

Although graphene aerosols and CNT macrostructures have high potential for very efficient SS-PCM fabrication, the production of these carbon supports is hampered by the use of harmful and expensive precursors or expensive equipment, which makes difficult their production at large scale. In this regard, *carbon aerogels* derived from glucose, cellulose, melons, potatoes, etc., represent an attractive alternative. These supporting materials exhibit interconnected porous structure, high specific surface area, low weight and reasonable thermal conductivity. They have been used by several authors to produce PCM composites that have proven excellent shape stability, high latent heat of melting (115–164 J/g) and thermal conductivity up to 0.94 W/m/K using paraffin waxes and PEG (Huang et al. 2019).

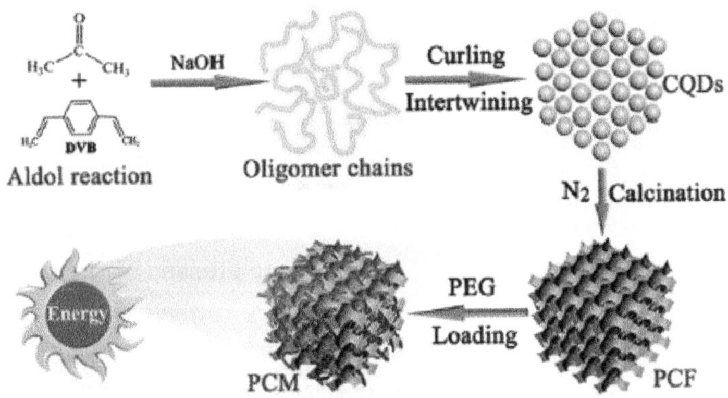

Figure 5.9. *Illustration of highly graphitized carbon foam preparation method and infiltration with PEG8000 (Chen et al. 2018). For a color version of this figure, see www.iste.co.uk/odru/heat1.zip*

Recently, Chen et al. (2018) reported a facile, low-cost and controllable method to produce a highly graphitized 3D carbon structure from acetone and divinylbenzene for PEG PCMs (Figure 5.9). At the maximum loading of 85 wt% PEG, the thermal conductivity of the SS-PCM is enhanced by 236% compared to pure PEG and no reduction in latent heat is observed despite the fact that the composite contains 15 wt% carbon.

5.3.2.4. *EG-based SS-PCMs*

Graphite is the most stable form of carbon under standard conditions. It is composed of graphene layers bonded by Van der Waals forces. EG is obtained by expansion or exfoliation of graphene via intercalation of volatile compounds between the graphene layers and rapid increase in temperature. As shown in Figure 5.10, this leads to a lightweight, highly porous honeycomb-like carbon structure with pore size in the micrometer range and high specific surface area.

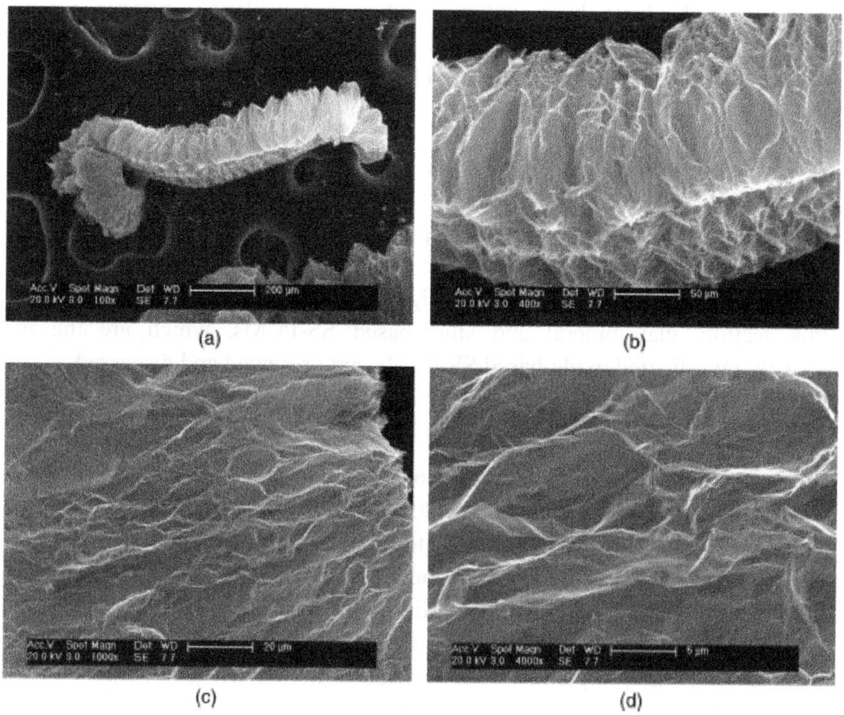

Figure 5.10. *SEM images of EG at different magnifications (Tao et al. 2015)*

Besides low density and large pore volume, other assets of EG are high thermal conductivity, high-temperature thermal stability and low cost. For these reasons, EG has been extensively investigated as support material for SS-PCM production. SS-PCMs with paraffin waxes, fatty acids and PEG have been successfully achieved with high PCM loading (often above 90 wt%), high latent heat (136-187 J/g) and thermal conductivity ranging from 0.6 to 5.3 W/m/K (Huang et al. 2019). An increasing attention is recently being paid to inorganic PCM/EG shape-stabilized composites. EG has proven to be efficient in encapsulating salt hydrates as well as molten salts, with PCM loading of 70–90 wt%, high latent heat of melting and significantly enhanced thermal conductivity (Huang et al. 2019). They have also proven to overcome segregation and phase separation issues that characterize salt hydrates. Moreover, SS-PCMs with very high thermal conductivity of 51 W/m/K and 31.5 W/m/K have been obtained when encapsulating eutectic mixtures KNO_3-$NaNO_3$ (Zhao et al. 2014) and $LiNO_3$-KCl-$NaNO_3$ (Xu et al. 2017), respectively, with 15 wt% and 25 wt% EG loading.

5.3.3. *Porous oxide-based SS-PCMs*

Besides porous carbon, various types of porous oxides have also been investigated as support for SS-PCMs due to their high thermal stability, chemical compatibility with many PCM families, high specific surface area and porosity, etc. In this section, clay-mineral and silica-based SS-PCMs, which are the most extensively investigated oxide-based SS-PCMs, are presented and discussed.

5.3.3.1. *Clay mineral-based SS-PCMs*

Most clay mineral materials have a porous structure with high specific surface area, which make them good candidates for SS-PCMs. Indeed, they can effectively load PCMs by capillary forces, surface tension, hydrogen bonds and/or Van der Waals force. Common porous clay mineral materials include kaolin, diatomite, sepiolite, montmorillonite, expanded perlite, attapulgite and expanded vermiculite (Table 5.7). All of them have high thermal stability and prove to be chemically compatible with many organic PCMs (e.g. paraffin waxes, fatty acids, PEG and alcohols) and some inorganic ones (e.g. salt hydrates and high-temperature nitrate salts). Moreover, they are abundant and low-cost materials. Recently, Lv et al. (2017) published a review on clay mineral-base SS-PCMs that we summarize here.

Clay mineral materials offer a wide variety of morphologies and microstructures, as can be seen in Figure 5.11. For example, montmorillonite occurs in platelets with inner layered microstructure, sepiolite has a fibrous structure, disk-type diatomite has quite ordered porosity and expanded perlite is highly porous and looks like honeycomb cages with very thin walls.

Clay mineral material	Short description
Kaolin	Clay rock contains chemical elements such as Si, Al, Fe, Ti, Ca, Mg, K, Na, etc. Silicon accounts for more than 50 wt%. Layered structure with small interlayer spacing and low cation exchange capacity compared with other clay mineral materials.
Diatomite	Siliceous rock mainly composed of silicon dioxide (up to 90 wt%). Highly porous material (up to 90%) with different possible morphologies and porous space characteristics.
Sepiolite	Silicate mineral with fibrous structure. Composed mainly of SiO_2 (approximately 52 wt%) and MgO (approximately 23 wt%).
Montmorillonite	Magnesium aluminum hydroxysilicate, with other possible elements. Layered structure with high cation exchange capacity.
Expanded perlite	Is a vitreous volcanic rock that has the unusual property of expanding tremendously when heated sufficiently. SiO_2 is the main constituent of perlite (up to 75 wt%).
Attapulgite	Magnesium-rich aluminum silicate rock with quite large specific surface area (300–600 m^2/g).
Expanded vermiculite	Mineral of the mica group formed by iron or magnesium silicates. Expands when heated sufficiently and has ion exchange capacity.

Table 5.7. *Short description of common porous clay mineral materials*

Clay mineral based SS-PCMs with organic PCMs are the most studied so far, although some inorganic PCMs have also been used (see Table 5.8). Commonly employed preparation methods are direct/vacuum infiltration, melt/solution intercalation, fusion absorption, etc. Before the incorporation of the PCM into the porous structure, the latter must be cleaned to remove impurities and often modified to activate the clay (e.g. production of expanded perlite, expanded vermiculite and expanded kaolin from raw perlite, vermiculite and kaolin). Acid/alkali treatment, high-temperature calcination, using intercalators, hydrothermal methods, etc., are commonly used to this end (Lv et al. 2017).

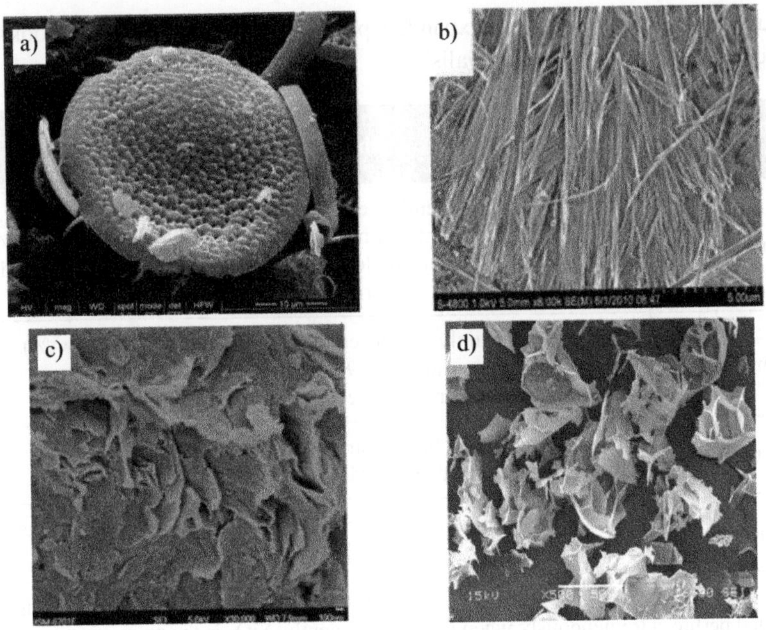

Figure 5.11. *SEM images of (A) calcined diatomite, (B) sepiolite (C) montmorillonite and (D) expanded perlite (source: Lv et al. 2017)*

Clay mineral support	PCM	PCM loading (wt%)
Kaolin	PW, FA, PEG	16–51
Diatomite	PW, FA, PEG, $LiNO_3$, Na_2SO_4	40–66
Sepiolite	PW, FA, alcohols	44–66
Montmorillonite	PW, FA, PEG, alcohols	21–86
Expanded perlite	PW, FA	25–87
Attapulgite	PW, FA	26–45
Expanded vermiculite	PW, FA	20–70

Table 5.8. *PCM loading ranges in studied porous clay mineral-based SS-PCMs (PW: paraffin waxes; FA: fatty acids; PEG: polyethylene glycol)*

As shown in Table 5.8, clay mineral-based SS-PCMs are characterized by low PCM loading and, therefore, low latent heat. Only in a few cases has PCM loading above 60 wt% been achieved (Lu et al. 2014). Moreover, the thermal conductivity of clay mineral-based SS-PCMs is quite low compared with carbon-based carriers. To improve their thermal conductivity several strategies have been adopted (Huang et al. 2019), such as pore surface modification or coating with high thermal conductive materials (e.g. decoration of the porous surface with silver nanoparticles, graphene coatings) or the incorporation of EG into the PCM.

The most suitable application for this group of SS-PCMs is passive heating and thermal comfort in buildings through their integration in construction materials.

5.3.3.2. *Silica-based SS-PCMs*

Porous silica materials have also been widely studied as supporting media to fabricate SS-PCMs. They are characterized by large specific surface area, high interconnected porosity and controllable pore structure. They also are fire-resistant, eco-friendly in nature and cost effective.

Porous silica materials can be divided into zeolites, mesoporous silica, silica nanoparticles, porous glass, aerogels and xerogels (Mitran et al. 2021). Zeolites possess an ordered structure and micropores less than 2 nm in diameter, which make them unsuitable for hosting PCMs. Silica particles are not porous but their spherical and monodisperse nature gives rise to interparticle porosity. Indeed, fused silica can be obtained from agglomeration of silica nanoparticles and used to obtain SS-PCMs. Silica aerogels and xerogels have a disordered and interconnected pore network and are characterized by low density (approximately 0.003 g/cc), very high specific surface area (up to 2,000 m^2/g) and very low thermal conductivity around 0.005 W/m/K. Porous glass has a similar pore structure, although with less porosity than aerogels; mesoporous silica nanomaterials (MSNs) are very versatile materials, whose structural properties can be easily tailored. They contain monodisperse mesopores in the 2–50 nm diameter range, which are usually ordered into hexagonal, lamellar or cubic arrangements. MSNs have thicker pore walls than aerogels and thus higher mechanical stability. Their specific surface area is in the 500–1,000 m^2/g range and total mesopore volume is up to 3.5 cm^3/g. Sol-gel synthesis is the most commonly used method to obtain porous silica nanomaterials.

The surface of porous silica nanomaterials is covered with silanol moieties that can participate in intermolecular hydrogen bonding with suitable encapsulated molecules. But their surface properties can be easily changed by introducing organic functional groups or by doping the silica framework with ad-atoms, thus modifying the wettability properties and offering the possibility of anchoring different types of molecules. Together with easy tailoring of the structural properties, this makes silica nanomaterials very versatile supporting materials.

There is a large variety of silica-based SS-PCMs that have been investigated. They differ in the PCM and porous silica nanomaterial used, as well as in the preparation method. The recent review article by Mitran et al. (2021) provides an excellent and comprehensive presentation of the work done to date on this category of SS-PCMs. We provide here a brief summary of the experience acquired and performances of silica-based SS-PCMs.

Paraffin waxes, n-alkanes, fatty acids and polyethylene glycol are the most frequently used PCMs in silica-based SS-PCMs, although some inorganic PCMs (e.g. salt hydrated and nitrate salts) were also employed. Some anhydrous inorganic salts, such as non-alkali molten salts and molten salts with high basicity (e.g. hydroxides and carbonates), react with silica at high temperature and cannot be used in combination with silica materials.

The silica-based SS-PCMs are usually obtained by either direct sol-gel synthesis of silica framework in the presence of PCM molecules or through the impregnation of molten PCMs or their solutions into porous silica matrices. The direct sol-gel synthesis is a simpler method, but obtained SS-PCMs are typically limited to 60–70 wt% PCM loading. In contrast, the impregnation method can achieve more than 70 wt% PCM loading while maintaining shape stability. Moreover, impregnation methods offer the possibility of tailoring the structural properties of the porous matrix, as well as the pores' surface chemistry, before the addition of the PCM.

			SS-PCM		
PCM class	T_m (°C)	ΔH_m (J/g)	wt% PCM	T_m (°C)	ΔH_m (J/g)
Paraffin wax n-alkanes	6.2–63.8	107–249	45–92	–7.4 to 63.7	95–156
Fatty acids	12–68.4	120–239	30–86	0.2–61.3	43–158
PEG	18.5–62.4	118–213	36–97	19.6–80	58–167
Hydrated salts	11.8–58.8	118–227	70–85	8.3–54.3	64–148
Molten salts	222–889	97–370	50–90	200–887	53–292

Table 5.9. *Silica-based SS-PCMs. Aggregated data from (Mitran et al. 2021): Tm: melting point, ΔHm: latent heat of melting, wt% PCM: PCM loading*

Table 5.9 provides some aggregate data of silica-based SS-PCMs for different families of PCMs. It shows that composite materials with high PCM loading (up to 92 wt%) and high latent heat of melting (>150 J/g) can be obtained. The maximum PCM loading allows more shape stability and, therefore, latent heat storage capacity is influenced by the nature of the PCM, the porous silica support and the preparation method:

– *PCM*: In general, the highest PCM loading is obtained with PGE, followed by fatty acids, while paraffin waxes lead to the lowest loading fractions. This is true for pristine silica matrices containing silanol groups on the surface, which favor the formation of a hydrogen bond between PGE or fatty acids and the pore's surface. Functionalization with hydrophobic organic substituents (e.g. alkyl groups) reverses the trend: highest loading fractions are obtained with paraffin waxes, while the lowest values correspond to PGE.

– *Silica support*: In general, supports with lower pore volume such as silica nanoparticles and fused silica present the lowest PCM loadings (<70 wt%). The highest values (>80 wt%) are obtained for high porosity supports such as aerogels and mesoporous silica. The pore size also influences the storage capacity. It is proven that there is a non-melting layer that forms on the surface of the silica pores, which remains liquid and does not contribute to latent heat storage. Below 30–40 nm, the relative weight of this non-melting layer compared to the total mass of PCM enclosed within the pores is significant and a strong reduction of the storage capacity can be observed.

– *Preparation method*: Generally, the best results in terms of PCM loading and heat storage capacity are obtained through impregnation of molten PCM, especially under vacuum. Solution synthesis leads to lower loading, but in general still higher than the direct sol-gel method.

Because silica nanomaterials have low thermal conductivity, most SS-PCMs are good insulators. However, various solutions could be applied if increased thermal conductivity is needed. Most of them are based on the addition of carbon materials, such as GO or EG, to the PCM or silica framework (Mitran et al. 2021). In such way, the thermal conductivity of the SS-PCM can be tailored between 0.1 and 2 W/m/K.

There are other specific effects due to nanoconfinement of PCMs into porous silica. The melting point of the PCM is often decreased with respect to the bulk and this effect can be larger than 10°C for mesopores. Supercooling is also influenced. While melting starts from the interface with the silica pores, crystallization starts from the volume and thus is kinetically hindered, resulting in amplified subcooling for molecules such as PEG or salt hydrates. On the contrary, segregation and phase separation issues characteristic of salt hydrates are overcome.

5.3.4. *Summary of porous supports and perspectives*

Main assets and disadvantages of the different porous supports used for SS-PCMs production are summarized in Table 5.10. Comparatively, carbon-based supporting materials are much more efficient in increasing the thermal conductivity of the PCM than oxide-based supporting materials. They also allow higher PCM loading fractions, thus leading to composites with generally higher latent heat storage capacity. Moreover, some of them exhibit efficient solar-to-thermal and electricity-to-thermal energy conversion. Whereas the production of graphene aerogels and CNT macrostructures is expensive, carbon aerogels and EG are cheap materials. The latter is likely the porous carbon support with the best performance.

	Supporting material	Main assets	Drawbacks
Porous carbon-based SS-PCMs	**Graphene aerogels**	– Lightweight – High porosity – Good thermal conductivity – Good solar-to-thermal and electricity-to-thermal energy conversion	– Production hampered by the use of harmful and expensive precursors or expensive equipment
	CNT macrostructures	– Lightweight, flexible – Highly porous – Good thermal conductivity – Good solar-to-thermal and electricity-to-thermal conversion	
	Carbon aerogels	– Low cost – Derived from natural resources – Super lightweight – High electrical conductivity	– Moderate thermal conductivity
	Expanded graphite	– Low cost – Low density – Large pore volume – High PCM loading – High thermal conductivity	
Porous oxide-based SS-PCMs	**Clay minerals**	– Low cost – High thermal stability – Eco-friendly – Fire resistant	– Low thermal conductivity – Low specific surface area – Low maximum PCM loading
	Silica	– Cost-effective – Large specific surface area – Fire-resistant – Very versatile materials: tunable pore, size pore surface chemistry	– Low thermal conductivity – Strong confinement effect

Table 5.10. *Summary of main assets and disadvantages of commonly used porous supports*

Oxide-based supporting materials are characterized by low thermal conductivity and moderate-to-low maximum PCM loading. However, they are eco-friendly, very low cost and very stable at high temperatures. Moreover, porous silica-based materials, especially mesoporous silica, are very versatile. Both the pore size and the surface chemistry can be easily tailored.

In spite of the tremendous progress achieved in SS-PCMs to date, there are still several issues to be addressed in future research:

– The advantages afforded by SS-PCMs come at the cost of a loss of energy storage capacity, generally on the order of the mass fraction of the support material. The design of new SS-PCMs that minimize this drawback, such as those recently proposed using PCMs with solid–solid transitions as support material (Serrano et al. 2021), would be desirable.

– The increase in thermal conductivity achieved so far is generally low to moderate, with some exceptions as EG/salt composites with 15–25 wt% of EG studied by Zhao et al. (2014) and Xu et al. (2017). The search for new supporting materials with very high thermal conductivity and high PCM loading capacity would be necessary. For instance, Ji et al. (2014) demonstrated that ultrathin-graphite foams (UGFs) could be employed to enhance the PCMs' thermal conductivity up to 18 times with only 1.2 vol% of supporting material.

– With rare exceptions, PCM subcooling is not improved. Moreover, it may even be amplified in relatively small pores. This is a major obstacle to the use of SS-PCMs and the problem should be solved.

– The effects of nanoconfinement on melting and crystallization are not yet well understood. It has been found that large shifts in transition temperatures can occur as pore size is reduced. Substantial increases in latent heat of fusion are also sometimes observed. A better understanding of the effect of pore size, pore morphology and pore surface chemistry on melting and crystallization would provide a useful tool for tailoring the phase change properties of the PCMs.

– Multifunctional SS-PCMs, such as those showing photothermal, electricity-to-thermal or magnetic-to-thermal energy conversion, are a very recent class of materials that deserves further investigation. They contribute to extend the application fields of PCMs.

It should also be noted that SS-PCMs have so far been produced on a laboratory scale, with no studies of their cost and production on an industrial scale. Studies on their durability are also scarce and, in many cases, the thermal characterization carried out is limited to the measurement of transition temperatures and enthalpies, which is insufficient for their evaluation.

5.4. Conclusion

PCM confinement into micro-/nanoscale frameworks results in novel materials able to face most of the problems encountered in solid–liquid PCMs. ME-/NE-PCMs and SS-PCMs have proven to be efficient in preventing liquid leakage and protecting the PCM against environmental influences. They are also able to enhance heat transfer by increasing either the surface-to-volume rate of the PCM or the apparent thermal conductivity or both. They also provide many other advantages such local management of volume changes, flammability reduction, mitigation of corrosion issues, overcoming phase separation and segregation phenomena, and easier handling and integration into the application.

In spite of the aforementioned advantages, ME-/NE-PCMs and SS-PCMs have some drawbacks such as increased cost of the storage material, loss of energy storage capacity and often supercooling amplification. Also, the thermal conductivity enhancement ratios achieved are very low compared to those obtained by composite materials based on metallic or graphitic foams.

Compared to ME-/NE-PCMs, SS-PCMs have several important advantages:

– Core–shell PCM encapsulation generally involves high preparation cost (ca. 50% of the end-material cost in ME-NE-PCMs with polymeric shell). On the contrary, SS-PCMs are usually prepared by simple and cheap methods and, therefore, it is expected that they result in much cheaper end-materials.

– Contrary to SS-PCMs, most of developed ME-/NE-PCMs were not able to improve significantly the thermal conductivity of the PCMs. Indeed, the limitation imposed to heat transfer by the thermal resistance of the shell usually counterbalances the positive effect of the large surface-to-volume ratio of the capsules as well as that of conductive nano-filler eventually added to the core.

– SS-PCMs also could solve the problems associated with micro- or nano-encapsulation technology, namely, the high number of parameters and type of preparation methods influencing the performance of the final material. It is worth noting that the preparation method of SS-PCMs is dependent on the organic or inorganic nature of the supporting material, but independent on the polymer or the inorganic framework used. In addition, melted PCM leakage when a capsule shell is damaged is another problem of ME-/NE-PCMs that can be avoided by using SS-PCMs.

Therefore, the attention paid to the development of SS-PCMs is growing. However, ME-/NE-PCMs remain an essential class of composite PCMs because they have better mixing properties with construction materials than SS-PCMs. In addition, PCM slurries cannot be formed using SS-PCMs.

Concerning the different classes of SS-PCMs, those using organic polymers as supporting material can be easily shaped and prepared with desired dimensions, whereas this is more complicated for those based on inorganic frameworks. On the contrary, the latter can be easily incorporated into construction materials. Among SS-PCMs with inorganic supports, those based on porous carbon materials are attracting increasing attention because they generally lead to superior performances. Indeed, their large pore volume, specific surface area and low density is expected to enhance the shape stabilization capability while minimizing the loss of energy storage capacity. Moreover, carbon-based supports have good thermal conductivity, thus enhancing the heat transfer ability of the PCM without additives. In addition, carbon-based materials can be used to encapsulate almost all types of PCMs due to their excellent chemical stability.

In spite of the tremendous progress achieved in ME-/NE-PCMs and SS-PCMs to date, there are still several issues to be addressed in future research (e.g. further increase in PCM loading and thermal conductivity, subcooling suppression, control of nanoconfinement effects on melting and crystallization and providing additional functions by incorporation of functional materials). It should also be noted that only few ME-/NE-PCMs are commercialized. Regarding SS-PCMs, they have so far been produced on a laboratory scale, with no studies of their cost and production on an industrial scale.

5.5. References

Aftab, W., Huang, X., Wu, W., Liang, Z., Mahmood, A., Zou, R. (2018). Nanoconfined phase change materials for thermal energy applications. *Energy Environ. Sci.*, 11(6), 1392–1424.

Alba-Simionesco, C., Coasne, B., Dosseh, G., Dudziak, G., Gubbins, K.E., Radhakrishnan, R., Sliwinska-Bartkowiak, M. (2006). Effects of confinement on freezing and melting. *J. Phys: Condens. Matter*, 18, R15–R68.

Arshad, A., Jabbal, M., Yan, Y., Darkwa, J. (2019). The micro-/nano-PCMs for thermal energy storage systems: A state of art review. *Int. J. Energy Research*, 43, 5572–5620.

Borreguero, A.M., Rodríguez, J.F., Valverde, J.L., Peijs, T. (2013). Characterization of rigid polyurethane foams containing microencapsulted phase change materials: Microcapsules type effect. *J. Appl. Polymer Sci.*, 128, 582–590.

Chen, L., Zou, R., Xia, W., Liu, Z., Shang, Y., Zhu, J., Wang, Y., Lin, J., Xia, D., Cao, A. (2012). Electro- and photodriven phase change composites based on wax-infiltrated carbon nanotube sponges. *ACS Nano*, 6, 10884–10892.

Chen, Z, Cao, L., Shan, F., Fang, G.Y. (2013). Preparation and characteristics of microencapsulated stearic acid as composite thermal energy storage material in buildings. *Energy and Buildings*, 62, 469–474.

Chen, X., Gao, H., Yang, M., Dong, W., Huang, X., Li, A., Dong, C., Wang, G. (2018). Highly graphitized 3D network carbon for shape-stabilized composite PCMs with superior thermal energy harvesting. *Nano Energy*, 49, 86–94.

Dheep, G.R. and Sreekumar, A. (2014). Influence of nanomaterials on properties of latent heat solar thermal energy storage materials – A review. *Energy Conversion and Management*, 83, 133–148.

Doshi, J. and Renerker, D.H. (1995). Electrospinning process and applications of electrospun fibers. *Journal of Electrostatics*, 35, 151.

Fan, L. and Khodadadi, J.M. (2011). Thermal conductivity enhancement of phase change materials for thermal energy storage: A review. *Renew. Sustain. Energy Rev.*, 15, 24–46.

Fang, G., Tang, F., Cao, L. (2014). Preparation, thermal properties and applications of shape-stabilized thermal energy storage materials. *Renew. Sustain. Energy Rev.*, 40, 237–259.

Fei, B., Lu, H.F., Qi, K.H., Shi, H.F., Liu, T.X., Li, X.Z., Xin, J.H. (2008). Multi-functional microcapsules produced by aerosol reaction. *J. Aerosol Sci.*, 39, 1089–1098.

Gui, X., Wei, J., Wang, K., Cao, A., Zhu, H., Jia, Y., Shu, Q., Wu, D. (2010). Carbon natotube sponges. *Adv. Mater.*, 22, 617–621.

Hassan, A., Laghari, M.S., Rashid, Y. (2016). Micro-encapsulated phase change materials: A review of encapsulation, safety and thermal characteristics. *Sustainability*, 8, 1046.

Hu, X., Wu, H., Liu, S., Gong, S., Du, Y., Li, X., Lu, X., Qu, J. (2022). Fabrication of organic shape-stabilized phase change material and its energy storage applications. *Eng. Sci.*, 17, 1–27.

Huang, X., Chen, X., Li, A., Atinafu, D., Gao, H., Dong, W., Wang, G. (2019). Shape-stabilized phase change materials based on porous supports for thermal energy storage applications. *Chemical Engineering Journal*, 356, 641–661.

Imran Hussain, S., Ameelia Roseline, A., Kalaiselvam, S. (2018). Bifunctional nanoencapsulated eutectic phase change material core with SiO_2/SnO_2 nanosphere shell for thermal and electrical energy storage. *Mater. Design*, 154, 291–301.

Ji, H., Sellan, D.P., Pettes, M.T., Kong, X., Ji, J., Shi, L., Ruoff, R.S. (2014). Enhanced thermal conductivity of phase change materials with ultrathin-graphite foams for thermal energy storage. *Energy Environ. Sci.*, 7, 1185–1192.

Kenisarin, M.M. and Kenisarin, K.M. (2012). Form-stable phase change materials for thermal energy storage. *Renew. Sustain. Energy Rev.*, 16(4), 1999–2040.

Khadiran, T., Hussein, M.Z., Zainal, Z., Rusli, R. (2015). Encapsulation techniques for organic phase change materials as thermal energy storage medium: A review. *Solar Energy Materials and Solar Cells*, 143, 78–98.

Khodadadi, J.M., Fan, L., Babaei, H. (2013). Thermal conductivity enhancement of nanostructure-based colloidal suspensions utilized as phase change materials for thermal energy storage: A review. *Renew. Sustain. Energy Rev.*, 24, 418–444.

Konuklu, Y., Ostry, M., Paksoy, H.O., Chavrat, O. (2015). Review on using microencapsulated phase change materials (PCM) in building applications. *Energy and Buildings*, 106, 134–155.

Li, F., Wang, X., Wu, D. (2015). Fabrication of multifunctional microcapsules containing n-eicosane core and zinc oxide shell for low-temperature energy storage, photocatalysis, and antibiosis. *Energy Conversion and Management*, 106, 873–885.

Li, G., Zhang, X., Wang, J., Fang, J. (2016). From anisotropic graphene aerogels to electron- and photon-driven phase change composites. *J. Mater. Chem. A*, 4, 17042–17049.

Li, J., Lu, W., Luo, Z., Zeng, Y. (2017). Synthesis and thermal properties of novel sodium nitrate microcapsules for high-temperature thermal energy storage. *Solar Energy Materials and Solar Cells*, 159, 440–446.

Li, C., Yu, H., Song, Y., Liang, H., Yan, X. (2019). Preparation and characterization of PMMA/TiO_2 hybrid shell microencapsulated PCMs for thermal energy storage. *Energy*, 167, 1031–1039.

Liu, Z., Zou, R., Lin, Z., Gui, X., Chen, R., Lin, J., Shang, Y., Cao, A. (2013). Tailoring carbon nanotube density for modulating electro-to-heat conversion in phase change composites. *Nano Lett.*, 13, 4028–4035.

Liu, C., Rao, Z., Zhao, J., Huo, Y., Li, Y. (2015). Review on nanoencapsulated phase change materials: Preparation, characterization and heat transfer enhancement. *Nano Energy*, 13, 814–826.

Liu, L., Su, D., Tang, Y., Fang, G. (2016). Thermal conductivity enhancement of phase change materials for thermal energy storage: A review. *Renew. Sustain. Energy Rev.*, 62, 305–317.

Liu, H., Wang, D., Wu, D., Ji, S. (2019a). Fabrication and applications of dual-responsive microencapsulated phase change material with enhanced solar energy-storage and solar photocatalytic effectiveness. *Solar Energy Materials and Solar Cells*, 193, 184–197.

Liu, H., Wang, X., Wu, D. (2019b). Innovative design of microencapsulated phase change materials for thermal energy storage and versatile applications: A review. *Sustainable Energy & Fuels*, 3(5), 1091–1149.

Lu, Z., Xu, B., Zhang, J., Zhu, Y., Sun, G., Li, Z. (2014). Preparation and characterization of expanded perlite/paraffin composite as form-stable phase change material. *Sol. Energy*, 108, 460–466.

Lv, P., Liu, C., Rao, Z. (2017). Review on clay mineral-based form-stable phase change materials: Preparation, characterization and applications. *Renew. Sustain. Energy Rev.*, 68, 707–726.

McCann, J.T., Marquez, M., Xia, Y. (2006). Melt coaxial electrospinning: A versatile method for the encapsulation of solid materials and manufacture of phase change nanofibers. *Nano Letters*, 6, 2868–2872.

Milián, Y.E., Gutiérrez, A., Grágeda, M., Ushak, S. (2017). A review on encapsulation techniques for inorganic phase change materials and the influence on their thermophysical properties. *Renew. Sustain. Energy Rev.*, 73, 983–999.

Mitran, R-A., Ionita, S., Lincu, D., Berger, D., Matei, C. (2021). A review of composite phase change materials based on porous silica nanomaterials for latent heat storage applications. *Molecules*, 26, 241.

Nelson, G. (2002). Application of microencapsulation in textiles. *Int. Journal of Pharmaceutics*, 242, 55–62.

Nomura, T., Zhu, C., Sheng, N., Saito, G., Akiyama, T. (2015). Microencapsulation of metal-based phase change material for high-temperature thermal energy storage. *Scientific Reports*, 5, 9117.

Palkovits, R., Althues, H., Rumpleker, A. (2005). Polymerization of w/o microemulsions for the preparation of transparent SiO_2/PMMA nanocomposites. *Langmuir*, 21, 6048–6053.

Peng, G., Dou, G., Hu, Y., Sun, Y., Chen, Z. (2020). Phase change material (PCM) microcapsules for thermal energy storage. *Advances in Polymer Technology*.

Serrano, A., Durán, M., Dauvergne, J.L., Doppiu, S., Palomo del Barrio, E. (2021). Tailored transition temperature plastic crystals with enhanced thermal energy storage capacity. *Solar Energy Material and Solar Cells*, 220, 110848.

Shchukina, E., Graham, M., Zheng, Z., Shchukin, D. (2018). Nanoencapsulation of phase change materials for advanced thermal energy storage systems. *Chem. Soc. Rev.*, 47(11), 4156–4175.

Song, Q., Li, Y., Xing, J., Hu, J.Y, Marcus, Y. (2007). Thermal stability of composite phase change materials microcapsules incorporated with silver nano-particles. *Polymer*, 48, 3317–3323.

Sun, K., Liu, H., Wang, D., Wu, D.Z. (2019). Innovative design of superhydrophobic thermal energy-storage materials by microencapsulation of n-docosane with nanostructured ZnO/SiO_2 shell. *Appl. Energy*, 237, 549–565.

Tao, S., Wei, S., Yulan, Y. (2015). Characterization of expanded graphite microstructure and fabrication of composite phase-change material for energy storage. *J. Mater. Civ. Eng.*, 27(4), 04014156-1.

Tay, N.H.S., Liu, M., Belusko, M., Bruno, F. (2017). Review on transportable phase change materials in thermal energy storage systems. *Renew. Sustain. Energy Rev.*, 75, 264–277.

Umair, M.M., Zhang, Y., Iqbal, K., Zhang, S., Tang, B. (2019). Novel strategies and supporting materials applied to shape-stabilize organic phase change materials for thermal energy storage–A review. *Applied Energy*, 235, 846–873.

Xu, T., Li, Y., Chen, J., Liu, J. (2017). Preparation and thermal energy storage properties of $LiNO_3$-KCl-$NaNO_3$/expanded graphite composite phase change material. *Sol. Energy Mater. Sol. Cells*, 169, 215–221.

Xu, Q., Liu, H., Wang, X.D., Wu, D.Z. (2018). Smart design and construction of nanoflake-like MnO2/SiO2 hierarchical microcapsules containing phase change material for *in-situ* thermal management of supercapacitors. *Energy Convers. Manage.*, 164, 311–328.

Yang, J., Li, X., Han, S., Zhang, Y., Min, P., Koratkar, N., Zu, Z-Z. (2016). Air-dried, high-density graphene hybrid aerogels for phase change composites with exceptional thermal conductivity and shape stability. *J. Mater. Chem. A*, 4, 18067–18074.

Zhang, S. and Wang, Z. (2018). Thermodynamics behavior of phase change latent heat materials in micro-/nanoconfined spaces for thermal storage and applications. *Renew. Sustain. Energy Rev.*, 82, 2319–2331.

Zhang, L., Li, R., Tang, B., Wang, P. (2016). Solar-thermal conversion and thermal energy storage of graphene foam-based composites. *Nanoscale*, 8(30), 14600–14607.

Zhang, L., Yang, W.B., Jiang, Z.N., He, F.F., Zhang, K., Fan, J.H., Wu, J.Y. (2017). Graphene oxide-modified microencapsulated phase change materials with high encapsulation capacity and enhanced leakage-prevention performance. *Appl. Energy*, 197, 354–363.

Zhao, Y.J., Wang, R.Z., Wang, I.W., Yu, N. (2014). Development of highly conductive KNO_3/$NaNO_3$ composite for TES (thermal energy storage). *Energy*, 70, 272–277.

Zhong, Y., Zou, M., Huang, F., Lin, T., Wan, D. (2013). Effect of graphene aerogel on thermal behavior of phase change materials for thermal management. *Sol. Energy Mater. Sol. Cells*, 113, 195–200.

6

Latent Heat Storage Systems: Concepts and Applications

Marie DUQUESNE[1] and Wahbi JOMAA[2]
[1] *LaSIE, CNRS, La Rochelle Université, La Rochelle, France*
[2] *I2M, CNRS, Université de Bordeaux, Talence, France*

6.1. Introduction

Renewable energy sources constitute an interesting reservoir, but their intermittence requires a better anticipation of needs and implementation of energy storage systems. The development of thermal energy storage systems (TESSs) is currently a technological challenge with major energy and economic stakes in the necessary ongoing energy transition.

This chapter focuses only on latent heat thermal energy storage (LH-TES) technologies integrating phase change materials (PCM) as storage materials.

Due to their high-energy density, their capacity to store at quasi constant temperatures and the diversity of available materials, latent heat storage system technologies are particularly attractive for many applications.

A wide range of technologies, materials and storage methods are needed in order to meet the needs of multiple and varied applications. The main objectives of the design/selection of an LH-TES system (LH-TESS) are as follows: decarbonization, the improvement of energy efficiency and the reliability of installations and reduction of the associated costs.

Heat and Cold Storage 1,
coordinated by Pierre ODRU, Elena PALOMO DEL BARRIO.

To reach these objectives, several aspects and technological challenges must be considered (Figure 6.1): those inherent to the available source (nature, temperature, power, etc.), those related to use (heat demand and range of temperatures, etc.) and those related to the system integrating PCM in which the energy will be stored.

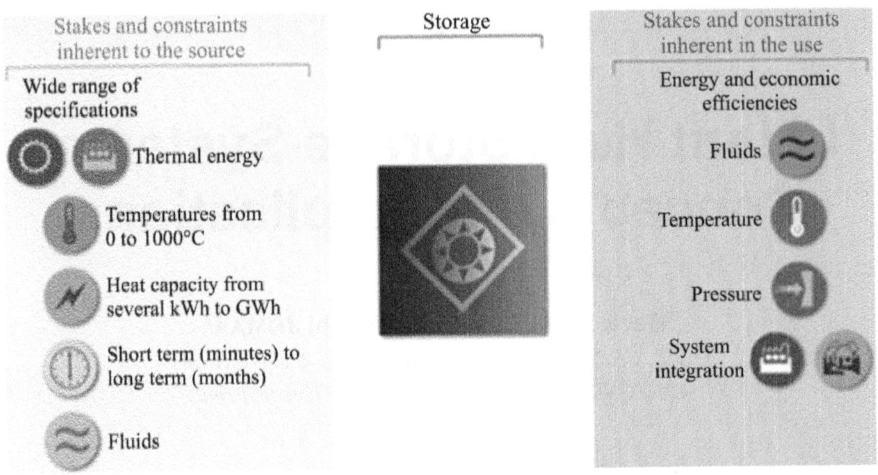

Figure 6.1. *Scientific and technological challenges related to TESS design (adapted from IEA ECES (2010))*

The cost of a TESS depends mainly on the following elements: PCM itself, the heat exchanger allowing the system energy charge and discharge and the space and/or chamber cost. LH-TESSs are assumed to be more competitive than sensible heat thermal energy storage systems (SH-TESSs) due to their capacity to store a large amount of heat at a constant temperature, their high-energy density, which reduces the volume of the storage tanks used, and their low cost. This leads to efficient storage in applications that include isothermal processes.

From a technical point of view, the most important design objectives are as follows: maximize compactness (storage capacity per unit volume), meet the charge/discharge power needs through quality heat transfer between the heat transfer fluid and the PCM, ensure the long-term mechanical and chemical stability of PCM (the system must resist to numerous charge/discharge cycles) and also compatibility between the heat transfer fluid, the heat exchanger and/or the PCM (security, durability), complete reversibility during a significant number of charge/discharge cycles (lifetime), small thermal losses and a certain ease of control (Gil et al. 2010).

The development of an efficient and economical storage technology therefore requires an optimized integration of PCM in TESS, then a designed TESS in the installations (Crespo et al. 2019) and also an adjustment of the connected components and of the operating parameters (tanks, heat exchanger and structural parts for charge and discharge, devices and components required for the operation and integration of TESS such as pumps, valves, control devices, etc.).

In this chapter, after a presentation of the types of systems, the focus will be on the main storage systems with high potential in terms of energy transition: cold storage, storage in solar power plants and in the fields of building and energy recycling in industrial processes. Other applications (textile, electronics, automotive, etc.) and the maturity of the mentioned systems will also be considered.

6.2. Types of systems and main components

The processes (heat generation and recovery systems, etc.) are often composed of a receiver, a heat transfer fluid and, possibly, a storage system. Storage systems performances are generally evaluated as a function of charge/discharge cycles whose definition may vary according to the fields of application. The system is considered to be charging when the PCM is collecting and discharging when the PCM is releasing. In the case of cold storage, the charge phase corresponds to a liquid/solid phase change, while in heat storage, the charge situation corresponds to solid/liquid phase change.

6.2.1. *LH-TESS classification*

Storage systems can be classified as active, passive or hybrid (Figure 6.2) (IEA ECES 2010). A system is considered active when the charge and discharge phases are controlled. A heat transfer fluid is then activated by an additional system (fan, pump, etc.). Conversely, the system is considered passive when the charge and discharge phases are underway and depend on environmental parameters. A system is considered hybrid if it combines these two storage modes.

Active systems are divided into direct and indirect systems. In a direct active system, the heat transfer fluid is also a storage material, where the storage material flows in the rest of the process. These systems can be found in sensible heat storage (SH-TESS), but seldom in LH-TESS except for the ice slurry and the slurries described below.

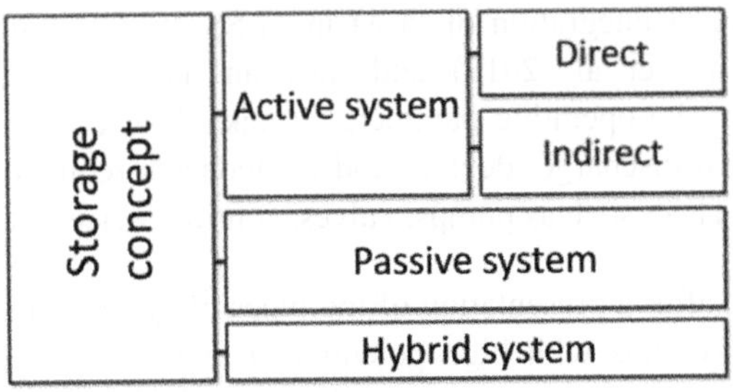

Figure 6.2. *Classification of various storage systems based on storage concept (adapted from IEA ECES (2010))*

In an indirect active system, the heat transfer fluid and the storage material are different. Such a system requires a heat exchanger allowing the heat transfer between PCM and the heat transfer fluid (Forster 2004; Herrmann et al. 2004; Laing et al. 2006; Gil et al. 2010; IEA ECES 2010; Kuravi et al. 2013; Hoffmann 2015; Barnes and Levine 2017). In the case of macro-encapsulated PCM, the storage system also plays the role of heat exchanger.

6.2.2. *Materials*

The operating temperature range of the installation/process guides the selection of the phase change material and of the heat transfer fluid (liquid, gas, or two-phase). This selection is essential, as it influences the definition of the heat exchange system (heat exchanger storage system with macro-encapsulation or additional heat exchanger). Therefore, it has an impact on the complexity of the associated technology, investment cost, etc.

6.2.2.1. *Phase change materials*

Phase change materials and their properties are approached in detail in Chapters 4 and 5. Their integration in the storage system must meet a certain number of economic, security or performance (thermal, physical, kinetic or chemical properties) criteria. These criteria are at the basis of the definition of the specifications that must be met by the selected materials:

– transition temperatures adapted to the concerned application and, for these temperatures, the highest possible transition enthalpies;

– high specific heat capacities and also high thermal conductivities to maximize the heat transfers and take advantage of the effects of sensible heat;

– high mass and energy densities to ensure the compactness of the storage systems;

– the closest possible densities in various phases in order to limit volume expansion during the phase change that may deteriorate the installation;

– phase change kinetics and a superfusion leading to power adapted to the needs of the application;

– lasting thermal and chemical stabilities so that the system can undergo a large number of charge/discharge cycles;

– the lowest danger in terms of toxicity, corrosiveness, explosiveness and flammability;

– cost compatible with industrial development.

The large families of PCM (Figure 6.3) and their properties are presented in Chapter 4. With the exception of metals, the thermal conductivity of PCM is low, which leads to performance problems during the operation of the storage system.

In the case of a liquid/solid transition, PCM crystallizes first in the regions near the walls of the heat exchanger and adheres to it. In this case, when heat is transferred by conduction, the low thermal conductivity becomes a significant drawback. The discharge process being ongoing, the solid PCM layer in contact with the exchanger develops and creates an increasing thermal resistance, therefore a decrease in the exchanged heat flux.

Various techniques have already been applied in order to improve the thermal conductivity of PCMs, such as the addition of nanoparticles, the use of metal foams or expanded graphite, micro or nano-encapsulation or shape-stabilization (see Chapter 5 for further details).

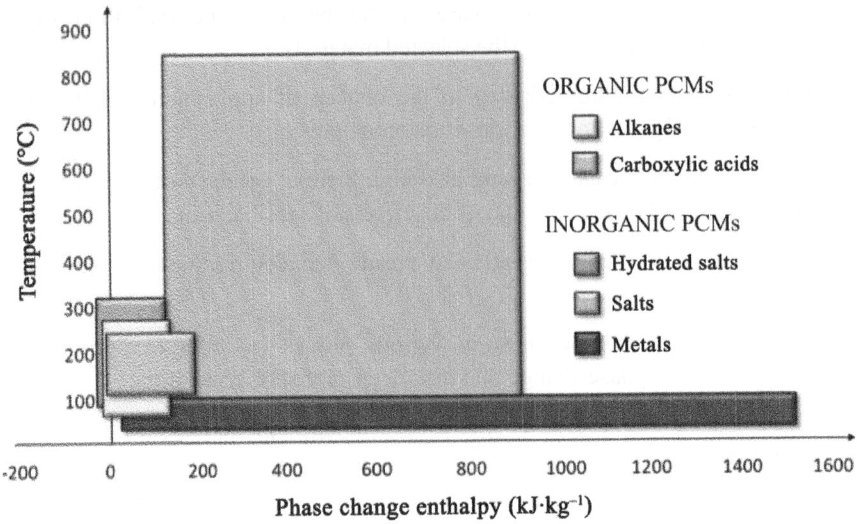

Figure 6.3. *PCM families according to temperatures and transition enthalpies. For a color version of this figure, see www.iste.co.uk/odru/heat1.zip*

6.2.2.2. *Heat transfer fluids*

The most commonly used heat transfer fluids in TESS are liquid water, molten salts, thermal oils and saturated vapor (the last requires an adapted installation able to resist high pressures). Liquid metals (liquid sodium) are no longer used as they pose risks and have a negative impact on the environment (Ferrière 2008).

Thermal oils are single-phase heat transfer fluids that are widely used due to their low corrosiveness. They allow reaching a maximal temperature of approximately 400°C. Highly flammable mineral oils have been replaced by synthetic thermal oils, which are the most commonly used in active-indirect SH/LH-TESS integrated in parabolic trough solar power plants (see Ferrière (2008)).

Recent works propose an innovative alternative: the replacement of synthetic oils with vegetable oils (thermal properties and operation comparable to that of a standard heat transfer fluid) and the use of many natural or recycled materials in the lining of an active-indirect storage system (thermocline; see Hoffmann 2015; Hoffmann et al. 2016). This association combines performance and availability of materials in industrial quantities, while reducing the environmental and financial impact.

Beyond 400°C, the heat transfer fluids used are mixtures of molten salts, water vapor and air. The majority of molten salts (mostly eutectic mixtures) used as heat transfer fluids allow reaching temperatures of about 550°C (which is the case of the Solar Two power plant; see Ferrière 2008; Gil et al. 2010; IEA ECES 2010). Most of them are sodium and potassium nitrates based. They have high-energy density and acceptable thermal conductivity.

Moreover, the experiments conducted in a direct active SH-TESS (the Thémis solar tower power; Solar Two, two tanks; Crespo et al. 2019) revealed an absence of steel corrosion and high stability of tested salts. However, they have a drawback that cannot be ignored. Due to their solidification at average temperatures ($T < 200°C$), their temperature must be maintained below 200°C when the installations are off, which imposes heavy and expensive constraints on the power plant operation (Ferrière 2008).

Liquid water is a competitive heat transfer fluid (low cost, no environmental impact, high conductivity and thermal capacity). However, at high temperature it imposes some complex technical issues related to piping resistance and tightness of circuits (when its temperature increases, high pressure is required to maintain water in a liquid state, for example, beyond 90 bars for 300°C).

6.3. Cold storage

Any system that maintains the temperature of a storage medium within a temperature range below 4°C is known as a cold thermal energy storage system (CTES) (Alva et al. 2018; Nazir et al. 2019). It integrates a liquid (cooled water) in the case of sensible heat storage or a PCM (ice, eutectic salts, etc.) in the case of latent heat storage. The main types of ice production and storage systems are represented in Figure 6.4 and compared in Table 6.1.

These systems emerged due to the needs for air cooling in individual buildings or in heat networks but also in many fields of application (food industry, pharmaceutical industry, etc.). The performance of such systems is difficult to evaluate in economic and energy terms, as it is influenced by many factors. Rismanchi et al. (2012) present a non-exhaustive study of the impact of various CTESS concepts from economic, energy and environmental perspectives. While the latter does not lead to general conclusions, it is obvious that the use of these systems has had positive effects overall by all accounts.

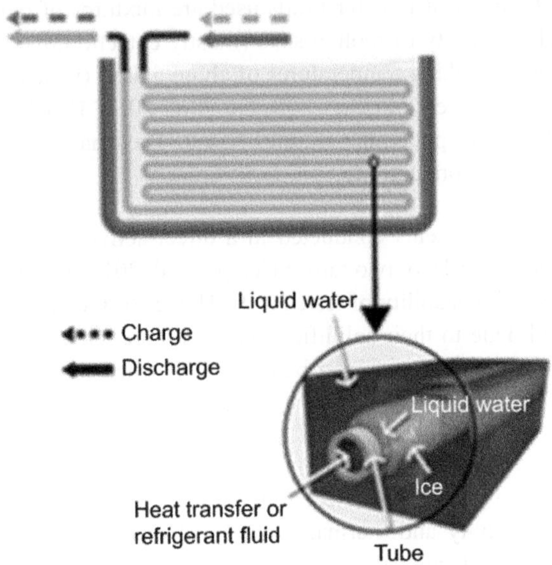

a)

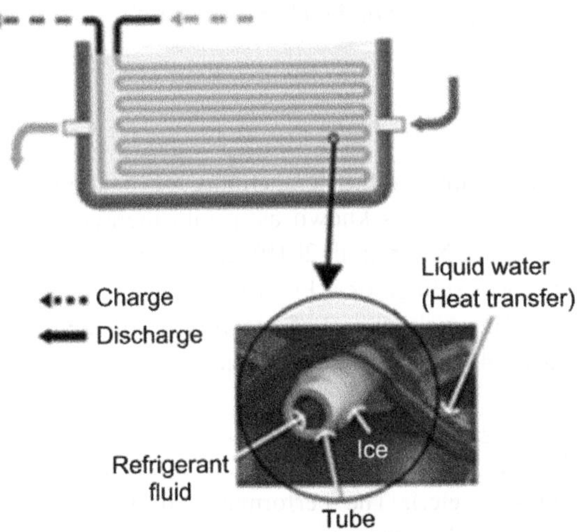

b)

Figure 6.4A. *Schematic representation of various types of cold storage (adapted from Wang and Kusumoto (2001)). a) Ice on internal melt tube; b) ice on external melt tube. For a color version of this figure, see www.iste.co.uk/odru/heat1.zip*

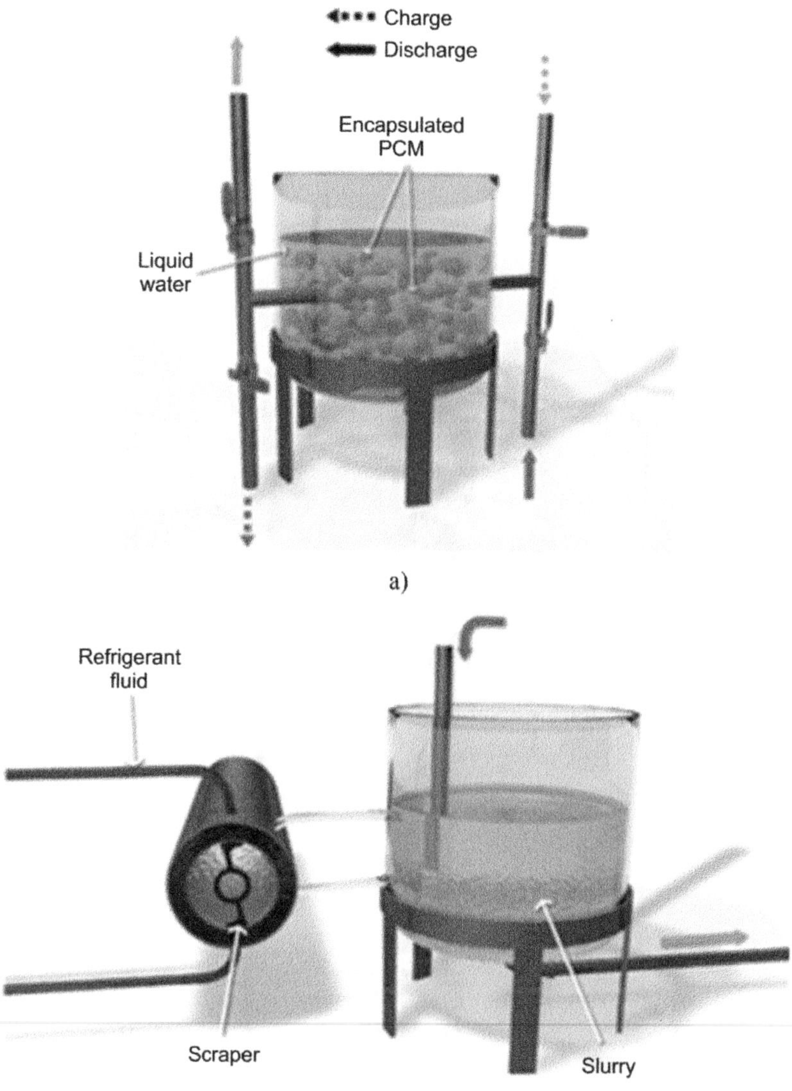

Figure 6.4B. *Schematic representation of various types of cold storage (adapted from Wang and Kusumoto (2001)) (continuation). a) Encapsulated; b) ice slurry. For a color version of this figure, see www.iste.co.uk/odru/heat1.zip*

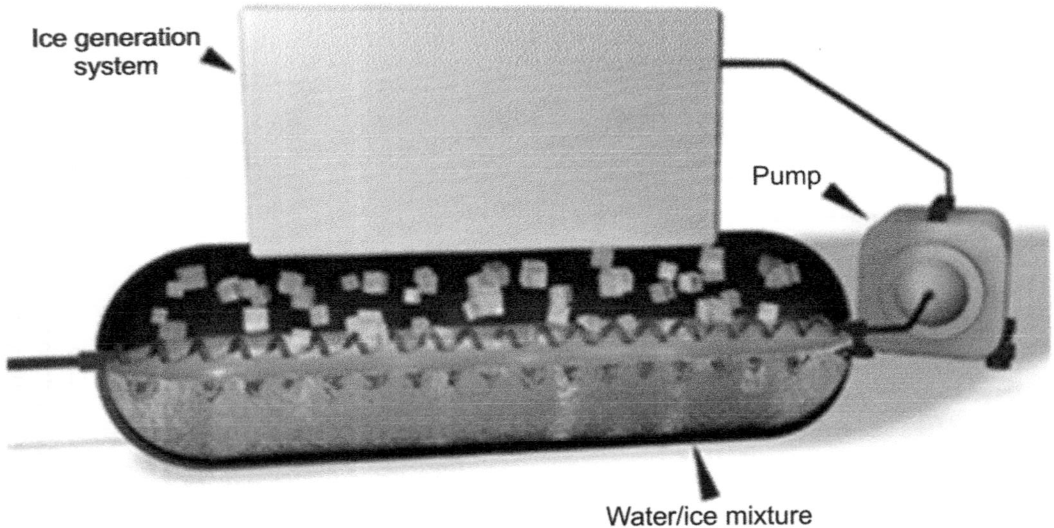

Figure 6.4C. *Schematic representation of various types of cold storage (adapted from Wang and Kusumoto (2001)) (continuation). Ice recovery. For a color version of this figure, see www.iste.co.uk/odru/heat1.zip*

1) Storage type		2) Ice on internal melt tube	3) Ice on external melt tube	4) Encapsulated	5) Ice slurry	6) Ice recovery
		7) Figure 6.4(A)(a)	8) Figure 6.4(A)(b)	9) Figure 6.4(B)(a)	10) Figure 6.4(B)(b)	11) Figure 6.4(C)
12) Storage concept		13) Indirect active	14) Direct active	15) Indirect active	16) Direct active	17) Direct active
18) Ice storage	19) Storage tank	20) Closed and pressurized chamber; requires resistance to pressure	21) Relatively high price; needs to be close to the cooler	22) Simple; not expensive; independent from the cooler	23) Compact; not expensive; independent from the cooler	24) Simple; not expensive; needs to be positioned below the cooler
	25) Operating mode	26) No clumping risk	27) Risk of ice clumping between the coils	28) Unequal charging if the capsules are initially partially charged	29) Simple	30) Potential ice clumping
31) System performance	32) Discharge temperature	33) Standard: 1–3°C 34) Max: 6–7°C 35) Stable increase with constant discharge rate	36) Standard: 1–2°C 37) Max: 5–6°C 38) Below 1.5°C until 80% of the ice melts	39) Standard: 1–3°C 40) Max: 7°C 41) Quasi constant until 80% of the ice melts	42) Standard: –3°C 43) Max: 2°C 44) Below 0°C until complete melting of the ice	45) Standard: 1–2°C 46) Max: 4–5°C 47) Relatively constant until 80–90% of the ice melts
	48) Response to charge variations	49) Hard to adapt	50) Hard to adapt, bubbling required	51) Hard to adapt	52) Easily adaptable, due to the significant exchange surface	53) Easily adaptable if ice does not get clumped

Table 6.1. *Comparison of various CTESS (adapted from Wang and Kusumoto (2001))*

6.3.1. *"Ice-on-tube" cold storage systems*

"Ice-on-tube" cold storage systems are the highest performance CTESS at present (Rismanchi et al. 2012). There are two versions: one qualified as "internal melt" and the other as "external melt". In both cases, a refrigerant fluid flows through an exchanger and exchanges with the storage medium consisting of a thermally insulated PCM tank. During the system charge phase (solidification), in either "internal melt" or "external melt" case, the temperature of the flowing refrigerant fluid is slightly below that of PCM solidification, which activates the phase change.

Upon discharge, two cases are identified. In the case of an internal melt CTESS (Figure 6.4(A)(a)), the heat transfer fluid is cooled by ice melting. The system is qualified as internal melt because ice melting starts from the surface of the tubes. This type of system is the most widespread and has the highest performance coefficients (Wang and Kusumoto 2001). In the case of an external melt CTESS (Figure 6.4(A)(b)), the liquid in the tank is set in motion and serves as heat transfer fluid. In this case, melting starts from the external surface of the ice formed on the tube.

A brief study of these two types of CTESS is proposed by Wang and Kusumoto (2001). This type of installation, both for the tank and the exchanger, has a relatively simple design allowing high performances at lower costs. It should however be noted that CTESS (tank) containing PCM in the case of external melt is slightly more complex due to the flow of liquid PCM. For this operation mode, the charge/discharge cycle can also be impacted in case of over-large clumping of solid PCM. This phenomenon is particularly highlighted by Torregrosa-Jaime et al. (2013), who show that the heat transfer occurring mainly through natural convection in the tank can progressively be replaced by conduction. The low thermal conductivity of PCM might thus affect the performances of the installation.

6.3.2. *Cold storage systems with encapsulated PCM*

This CTESS is particularly attractive due to the simplicity of its design and implementation and also to its very high performance during use (Wang and Kusumoto 2001). The PCM is encapsulated in envelopes of various natures and forms and clumped in a tank through which a heat transfer fluid flows (Figure 6.4(B)(a)).

Similar to the "ice on tube" systems, during the charge, the PCM contained in capsules crystallizes upon contact with the heat transfer fluid flowing at a temperature below that of PCM solidification. During the discharge, the PCM participates in the cooling of the heat transfer fluid during the endothermal melting reaction and can then be injected in the conditioning installation.

As already mentioned, this concept has attracted particular interest, and many studies have been conducted focusing on the parameters influencing the performances of this method (Chen et al. 2000) concerning both the capsules (size, material and nucleating agents) and the heat transfer fluid (flow rate and temperature).

Anecdotally speaking, it is also interesting to note that the emergence of micro-encapsulation techniques (Yamagishi et al. 1999; Hawlader et al. 2003) contributed to this enthusiasm though it only applies to PCMs whose economic viability for CTESS applications is still under study.

6.3.3. *Cold storage systems with ice slurry*

These direct systems with ice slurry are a type of dynamic storage system in which a two-phase (solid-liquid) medium qualified as slurry acts simultaneously as storage material and heat transfer fluid. The latter is generally a PCM mixed with a solution lowering the solidification temperature of PCM.

The system is based on a double tube heat exchanger in which a refrigerant fluid at a temperature below that of PCM solidification flows at the periphery while the slurry flows inside (Figure 6.4(B)(b)). The advantage of this system is the standard dissociation between the charge and discharge zones of the system. The charge (crystallization) takes place at the interface between the two tubes; the formed PCM crystals are then harvested and mixed with the slurry. The slurry, stored in a tank, is sent back to the conditioning system, where discharge takes place. This type of system has a relatively complex design and is characterized by quite low efficiency compared to the two previously mentioned types of systems (Wang and Kusumoto 2001).

Many studies focus on improving the overall efficiency of this type of system by evaluating the influence of various mixtures (Matsumoto et al. 2004) or of the exchange zone (Wijeysundera et al. 2004). However, the main difficulty of this method lies in its capacity to generate this two-phase medium of "slurry" in a continuous and efficient manner over a long period (Wang and Kusumoto 2001; Alva et al. 2018).

6.3.4. *Cold storage systems with ice recovery*

There are few references and studies on this system (Figure 6.4(C)) due to its relatively high cost and to its highly complex design (Knebel 1995; Wang and Kusumoto 2001).

The system consists of separating the production of ice and the storage of water/ice mixture. The latter is contained in a tank and is maintained at the PCM melting temperature by continuous phase change. If the ice layer at the surface of the exchanger becomes too thick, defrosting is required.

6.4. Applications in the building sector

Energy storage in a building reduces the fluctuations of internal temperature in relation to the variations of environmental temperature. It can be implemented by sensible heat by varying the temperature of the envelopes of the buildings, for example, or by latent heat for heating and cooling as in a passive system, but also in active systems (Mehling and Cabeza 2008).

The use of a latent heat storage system allows the stabilization of ambient temperature: when the latter is below the solidification temperature, the PCM releases heat (discharge phase) and, conversely, it stores heat when the temperature is higher than the melting temperature (charge phase). The major advantage of latent heat storage is a high-energy density for a small temperature difference between heat storage and release (Sharma et al. 2015).

6.4.1. *Passive systems*

The PCM can be integrated in the building as a passive component in practically all the elements of the envelope (Sawadogo et al. 2021). Nevertheless, they are most commonly integrated in walls, floors, ceilings, roofs and windows (Geetha and Velraj 2012; Kosny 2015a, 2015b).

Since 1996, Hamdam and Elwerr (1996) have worked on the modeling of latent heat thermal storage in a vertical PCM plate on a wall with the calculation of the evolution of the position of the liquid/solid interface in the material. This strategy has been the object of many publications, which are listed by several authors (Wang et al. 2009; Kuznik et al. 2011; Akeiber et al. 2016; Madad et al. 2018; Faraj et al. 2019). According to the conclusions of many studies on this process, the activation of thermal exchanges between air and PCM by forced convection would improve the storage/discharge performances.

Once the PCMs are selected, mainly based on the ranges of application temperatures and on their thermo-physical properties, it is important to examine their incorporation in the building materials or elements. There are four modes of incorporation that are often examined (Kuznik et al. 2011):

– *Mixture*: this may be the best cost-saving method, as it requires very little equipment. A liquid or powder phase change material is simply mixed with building materials during manufacturing. The main drawback of this technique is the risk of molten PCM leakage over time and also the degradation of the mechanical properties of the building material (Memon 2014).

– *Impregnation*: according to this method, the porous building material is impregnated with molten PCM, which will be soaked in by capillary effect and trapped inside the pores of the building material (Kaasinen 1992; Soares et al. 2013). One of the advantages of this method is the possibility to transform a standard building material into a phase change material, but it also runs the risk of losing PCM over time.

– *Shape stabilization*: this technique stabilizes PCMs and avoids leakage problems thanks to the interactions between the porous matrix and the PCM confined in the nanopores of the building material (Sayyar et al. 2014; Gao et al. 2018). There is a wide variety of materials used for PCM shape stabilization, notably polymers, porous materials and nanomaterials. Umair et al. (2019) have classified these materials into five large groups: micro-encapsulation, polymer matrices, nanomaterials, porous materials and solid–solid PCM.

– *Encapsulation*: in order to avoid PCM leakage risks, it is possible to encapsulate before incorporation. This is referred to as nano-encapsulation when the characteristic size is below 1 µm, as micro-encapsulation if this size is below 1 mm to 1 cm, and as macro-encapsulation when it exceeds the mm or the cm (Alehosseini and Jafari 2019). In the case of macro-encapsulation, the low thermal conductivity of PCMs may limit transfers (Kośn 2015).

PCM integration in the building envelope is the most commonly used among passive uses of PCM in buildings. It is thus possible to combine heat storage and building element functions. In this case, the most widespread structures are as follows:

– *PCM-plasterboards*: they generally consist of a light building material and PCM. The most commonly used PCMs are fatty acids, capric acids, lauric (palmitic and stearic) acids, butyl stearate, dodecanol and polyethylene glycol 600 (Tyagi and Buddhi 2007). PCM-plasterboards are generally fabricated either by immersion of the plasterboard in PCM or by PCM addition during the manufacturing of the plasterboard. These PCM-plasterboards are fixed on the walls as a sandwich composed of an insulator, polystyrene and PCM-plaster.

– *PCM-concrete blocks*: several experimental studies show the interest in incorporating PCMs in building elements of the concrete block type (Hawes et al.

1993; Kośny 2015; D'Alessandro et al. 2018; Shafigh et al. 2018; Adesina 2019; Arivazhagan et al. 2019; Berardi and Gallardo 2019). In this case, impregnation and mixture seem to be the most practical means for PCM incorporation. The two techniques have been used with satisfactory results. Several authors prefer the incorporation of organic PCMs in order to avoid the large volume variations during the phase change (D'Alessandro et al. 2018) while accepting the risks of leakage of liquefied PCM.

– *PCM-based Trombe wall*: it is an interesting application of inertia increase following the incorporation of PCMs. As an extension of the initial Trombe wall (Trombe 1974), they are composed of a regular masonry wall containing PCMs spaced at about 20 cm by a glass layer or a plastic glazing (Figure 6.5) with or without air flow between the masonry and the glazing (Bourdeau 1982; Delgado et al. 2019). Many experimental studies and numerical simulations focused on the choice of materials and the optimization of system dimensioning (Omara et Abuelnuo 2019). Experiments were conducted with various PCMs in order to have the highest thermal efficiency of a PCM-based Trombe wall: Benard et al. (1982) used paraffin wax, Farouk and Guceri (Farouk and Guceri 1981) used a mixture of Glauber salt and sunoco P-116 wax. According to these studies, the PCM-based Trombe walls require less space between the wall and the glazing than the standard Trombe walls. Moreover, they are lighter than the classic Trombe walls (Fiorito 2012).

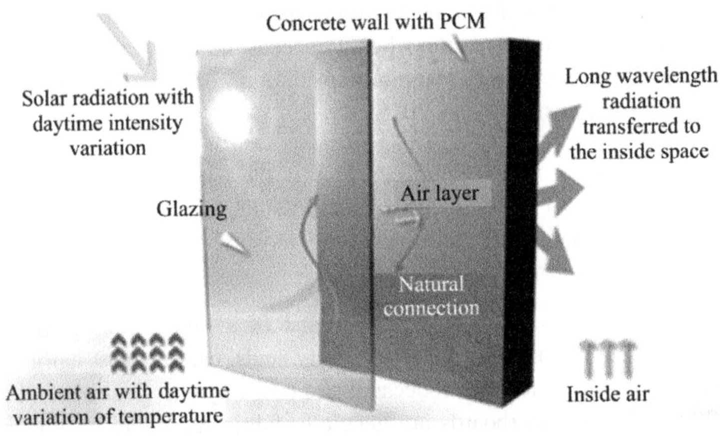

Figure 6.5. *Configuration of a Trombe wall with PCM (adapted from Omara and Abuelnuo (2019))*

6.4.2. *Active use of PCM in buildings*

In active systems the flow of fluid in the components is activated by a mechanical system (fan, pump, etc.). Due to this active character, the energy storage/discharge capacity can be used upon demand.

Active systems are generally composed of three elements:

– a "heat exchanger storage system" of latent heat thermal energy containing PCMs. It is the central element of the storage device;

– a flow circuit of the heat-transfer fluid (often air, sometimes water);

– a fan or a pump that sets the flow rate of the fluid in the exchanger.

Active systems of the heat exchanger type are used in various configurations for heating, domestic hot water and thermal comfort. For example, for the latter, they improve comfort inside buildings by circulating cool air (cooling) or hot air (heating) in the buildings, upon demand. If cooling is needed, during the day, the "hot" air in the building is circulated in an active system. The latter contains solidified PCMs. Being subjected to a temperature above their melting point, PCMs melt and absorb heat from the air. Once cooled, the air is reinjected in the building. To solidify the PCM, during the night, the air outside the building, whose temperature is below the PCM melting point, is circulated (Figure 6.6).

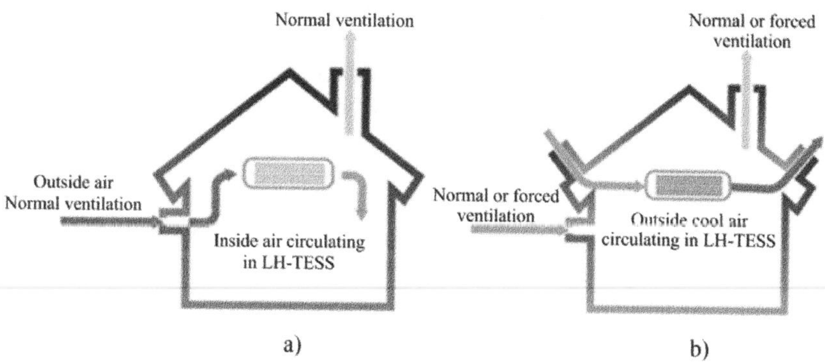

Figure 6.6. *Operating principle of the system with storage heat exchanger: (a) daytime operation; (b) night operation (adapted from Hed and Bellander (2006)). For a color version of this figure, see www.iste.co.uk/odru/heat1.zip*

Several works have focused on the evaluation of technical and economic performances when using such systems (Chaiyat 2015; Rastogi et al. 2015; Flores et al. 2017; Charvát et al. 2019; Faraj et al. 2021; Iten 2021; Javier Batlles et al. 2021; Mazzoni et al. 2021; Said and Hassan 2021). These performances are first related to the choice of the phase change temperature of PCM, which must be adequate to the weather conditions and at the level of the thermal conductivity of the chosen PCM. Most studies show a potential for reducing energy consumption of about 10–20% and recommend having an air (which is here the heat transfer fluid) flow rate during the charge phase three to four times higher than during the phase, in order to guarantee PCM solidification and optimal storage.

In the case of domestic hot water or water used for heating the building, water is an excellent material for sensible heat storage, as it is cheap, stable, renewable, clean, integrated into widespread and commercialized systems. However, its drawback is low-energy density (kWh/m^3), which results in significant storage volumes.

Latent heat storage technologies, based on the use of phase change materials, offer a possibility to avoid this drawback. It is then possible to use PCM instead of water as a storage material or to integrate macro-encapsulated PCM to a water tank (Figure 6.7), thus creating a hybrid water/PCM tank (Mehling and Cabeza 2008). When a hot water tank is associated with a PCM, the amount of energy stored in the tank increases (two to three times) and allows in particular the association with a solar collector and/or a reduction of the volume of the storage tank.

Experimentations and numerical simulations of such a system show a beneficial effect on stratification and also on the duration over which the temperature of hot water is maintained (Mehling et al. 2003; Cabeza et al. 2006; Talmatsky and Kribus 2008; Mazman et al. 2009; Fazilati and Alemrajabi 2013; Nkwetta et al. 2014; Najafian et al. 2015; Xie et al. 2017; Bouhal et al. 2018; Kılıçkap et al. 2018; Shirinbakhsh et al. 2018; Arivazhagan et al. 2019).

Moreover, several experimental and numerical studies have focused on the geometries of exchangers (tube exchangers, plate exchangers or cluster exchangers) and also on their location in the building. All these studies show that the choice of the exchanger and of the phase change material depends on the climate of the region and on the ranges of environmental temperatures (Rouault et al. 2013, 2016; Ortega Del Rosario et al. 2016; Mourid and El Alami 2017; Liu et al. 2018; Memarian et al. 2018; Pop et al. 2018; Zeinelabdein et al. 2018, 2019; Charvát et al. 2019; Dardir et al. 2019a, 2019b).

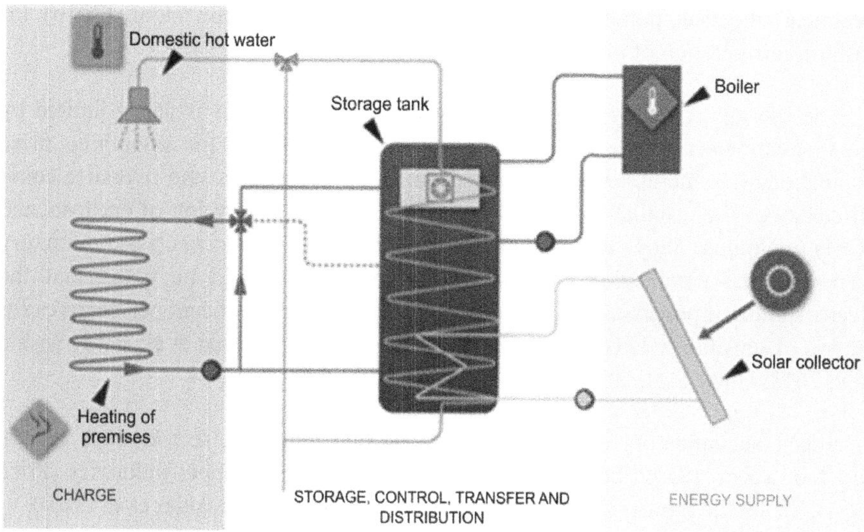

Figure 6.7. *Schematic representation of a combined water heating system with hybrid storage tank. For a color version of this figure, see www.iste.co.uk/odru/heat1.zip*

6.5. Applications in industry

Excess heat generated in industry is still dissipated into the environment. This excess heat, as shown by Miró et al. (2015), represents very significant quantities of heat that may be of the order of a quarter of the total annually generated energy. Using this "resource" thus allows both a reduction of CO_2 emissions and the reinvestment of this energy in order to reach higher energy efficiency. There are many works on the use of LH-TES in industry. Among the considered solutions, many designs use TESS and they depend mainly on the "nature" of this excess heat that can be globally referred to a level of temperature. An economic and technological study on the recovery solutions used in the industrial sector was conducted by Brückner et al. (2015).

Phase change cycles may produce cold during the periods when electricity is cheap or available in large amounts and store it for later use, thus being an attractive means for storing electricity (see Chapter 4 for further details).

There are storage systems using PCMs in the agri-food sector for the stages of transformation, storage or packaging. For example, the works of Devahastin and Pitaksuriyarat (2006) present a concept of solar drying using paraffin for the treatment of sweet potatoes or the systems mentioned by Lu et al. (2010) for uniform refrigeration of products in supermarkets.

The energy efficiency of the current waste recycling plants is partly limited by the fluctuation of combustion gas temperatures. A solution for smoothing these fluctuations has therefore been proposed in order to improve the overall electric efficiency of the installation (Hu 2019). It relies on the integration of encapsulated PCMs (aluminum alloys encapsulated in ceramic bricks similar to classic refractory bricks) in the combustion chamber. Furthermore, it should be noted that the electrification of processes and particularly of heat opens significant perspectives for the use of sensible or latent heat energy storage systems (Heinen et al. 2018; Seshie et al. 2018; Hutty et al. 2020).

Steam accumulators are also used in various industrial sectors such as the agri-food sector (Sardeshpande and Pillai 2012), glass and rubber industries (Price 1982), the textile industry (Goldstern 1979) and metallurgy (Shnaider et al. 2010).

More recently, steam accumulators are used in electric power plants, and particularly in solar thermal power plants (Steinmann and Eck 2006). The direct storage of saturated or superheated steam in pressurized systems is not efficient due to the low-energy density. To address this problem, steam accumulators, indirect storage systems, are used as buffer storage devices to store the excess saturated steam produced during peak periods for later use, thus allowing an extended operation of the industrial installation (Stevanovic et al. 2012; Sabihuddin et al. 2015).

During the discharge of such systems, pressure in the steam accumulator drops. To avoid this pressure drop, techniques such as adding a flash evaporator or encapsulated PCMs inside the liquid volume are considered. The advantage of adding encapsulated PCMs is that it avoids thermo-mechanical stresses resulting from phase transitions and accessing a significant storage capacity (of about 100 kWh/m^3 compared to that of water 20–30 kWh/m^3). Rigid metal or ceramics-based capsules are commonly used in high temperature applications such as the production of energy (Castellon et al. 2011; Liu et al. 2012; Jacob and Bruno 2015). Cylindrical capsules are commonly used in commercial steam accumulators (Medrano et al. 2010).

Several experimental and numerical studies have focused on the characteristics of LH-TES related to these cylindrical capsules during the thermal energy charge and discharge phases (Steinmann and Tamme 2008; Bourdillon 2015; Appasaheb 2020). There are works describing the optimization of PCM encapsulation in order to improve the charge and discharge time of LH-TES. For example, in Niyas (2018), the authors conduct a numerical study of a new technique for the improvement of heat transfer consisting in a tube-in-tube LH-TESS with a small radius for an efficient heat transfer in cylindrical capsules. Thanks to this technique, fewer capsules can provide a similar or higher storage capacity with comparatively shorter charge and discharge durations.

Moreover, in the case of nuclear power plants, the European Utility Requirements (EUR) requires the improvement of performances. This may be achieved through the integration of sensible or latent heat thermal energy storage systems at the outlet of the exchanger (Gomez 2011; Alameri 2015; Coleman et al. 2017; Alva et al. 2018).

Cogeneration is also an ascending technology for generating both electricity and heat. Furthermore, it allows a conversion at temperatures that may reach 600°C. However, since the needs for electricity and heat are not synchronous, a part of the heat that is continuously produced in cogeneration power plants is lost (Kavvadias et al. 2018). Up to 90% of the heat produced may be used by integrating a storage system. Kurata et al. (2014a, 2014b, 2014c) validated the interest of integrating a latent heat storage system in a 30 kW cogeneration unit at the University of Sapporo in Japan. At a larger scale, this integration is being developed at a cogeneration power plant in Sarre (Germany). This power plant comprises a gas turbine with a heat recovery steam generator. The thermal energy storage system is integrated between the steam generator and the water supply (Johnson et al. 2015, 2017; Gibb et al. 2018).

6.6. Concentrated solar power plants

Concentrated solar power (CSP) plants aim to transform the energy of solar radiation into heat to convert it into electric power by means of a thermodynamic power cycle coupled with a power generator (see Chapter 8 for further details). They are installed in geographical areas with strong insolation and allow the easier smoothing of electric power generation during the day due to a thermal energy storage that is cheaper than the battery systems.

Thermal storages at average high temperatures (above 120°C) for applications to CSP power plants have been widely studied (Crespo et al. 2019). The main advantages of their integration in solar power plants are as follows (Gibb et al. 2018):

– increase in power plant capacity (generation during the night, buffer during meteorological events leading to the absence of sunlight);

– improvement of energy efficiency (fewer cut-offs during the periods of strong generation, reduction of losses at turbine start-up);

– improvement of grid flexibility;

– cost reduction (maximized production at the peak of demand, smaller size of the solar field).

A CSP plant consists of the elements illustrated in Figure 6.8 (ADEME 2011):

– an optical device for concentrating solar radiation with a parabolic trough or a parabolic concentrator, a linear Fresnel concentrator or a tower concentrator (Ferrière 2008; Gil et al. 2010; Barnes and Levine 2017);

– a heat generation system composed of a receiver, a heat transfer fluid and, possibly, a thermal energy storage system, which in most cases consists of sensible heat thermal energy storage systems;

– a heat into electricity conversion subsystem.

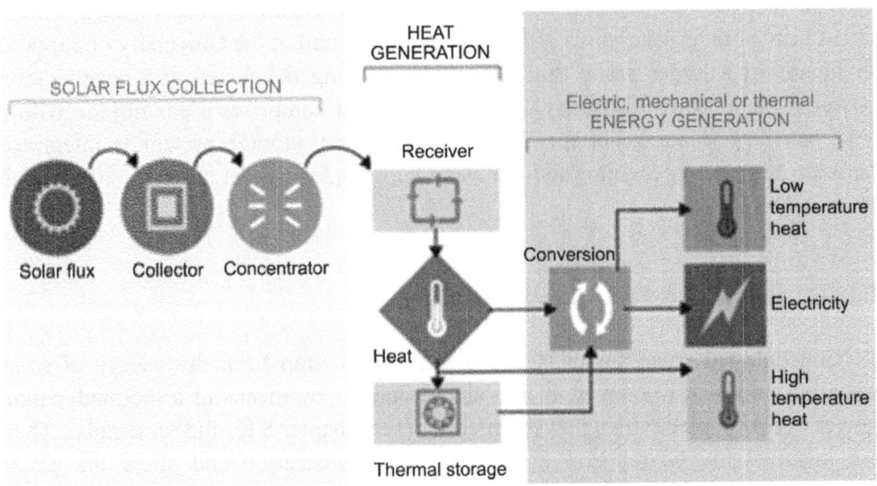

Figure 6.8. *Schematic representation of a CSP plant integrating a LH-TESS (adapted from ADEME (2011))*

Various sensible heat thermal energy storage systems (SH-TESS) have been designed for integration into CSP plants (Gil et al. 2010; Medrano et al. 2010; Pelay et al. 2017).

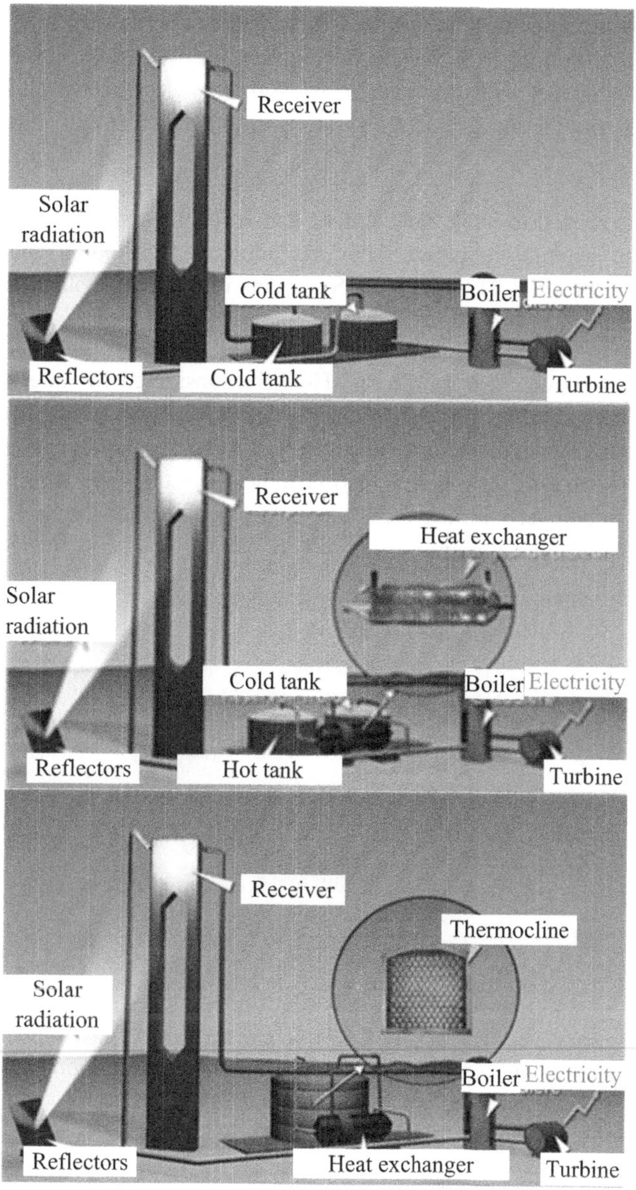

Figure 6.9. *Storage systems integrated in CSP plants (adapted from Pelay et al. (2017) and Alva et al. (2018)). (a) Direct active system with two tanks; (b) indirect active system with two tanks; (c) indirect active system with one tank. For a color version of this figure, see www.iste.co.uk/odru/heat1.zip*

Figure 6.9 presents the three most widely spread sensible heat storage systems: the direct active system with two tanks (Figure 6.9(a)) (Medrano et al. 2010), the indirect active system with two tanks (Figure 6.9(b)) (Kearney et al. 2003, 2004; Ferrière 2008) and the indirect active system with one tank (Figure 6.9(c)) (Pacheco et al. 2002).

The works related to latent heat storage as part of the application to CSP plants are still in an exploratory phase. They are aimed at developing and integrating LH-TESS or hybrid SH-TESS/LH-TESS systems into the CSP plants based on the existing systems.

Figure 6.10 shows an example of LH-TESS considered for integration into a CSP plant. This system uses the stored heat to generate steam through turbines. The heat is transferred from the solar collector to LH-TESS through a heat transfer fluid (generally a thermal oil or molten salts). The heat transfer fluid flows through the finned tubes in contact with the PCM to transfer and recover the heat. The heat is stored at phase change temperature.

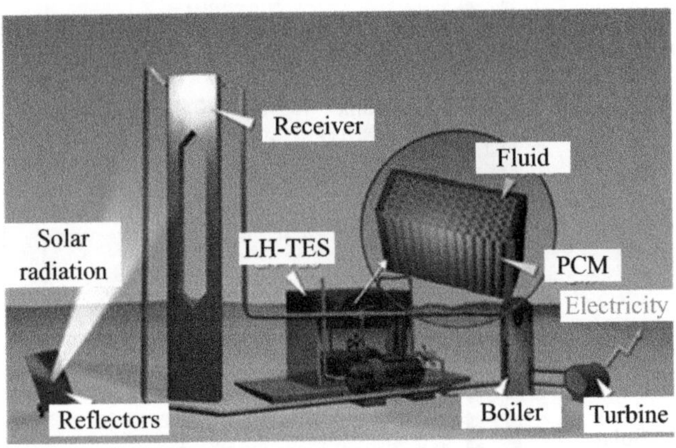

Figure 6.10. *Typical LH-TESS that can be used in a CSP plant (adapted from Pelay et al. (2017)). For a color version of this figure, see www.iste.co.uk/odru/heat1.zip*

Furthermore, the solar tower power plant PS10 (direct active SH-TESS, direct steam generation (IEA ECES 2010)) relies on the use of a central receiver with a saturated steam cycle at 250°C and 40 bars. This sensible heat storage concept with steam accumulators could be combined with or converted into LH-TESS with PCMs (Kuravi et al. 2013).

Indeed, in order to increase the storage capacity, a proposal was made to use LH-TESS jointly with the steam accumulator by placing PCMs directly inside or outside of the steam accumulator. They can also be used in indirect active storage configuration, in which a separate heat transfer fluid (such as synthetic oil) flows into a heat exchanger that is in contact with PCM (Kuravi et al. 2013).

A hybrid (indirect active LH-TESS and SH-TESS) system has also been proposed by DLR-ZSW in an application for CSP plants (IEA ECES 2010). The first prototype of a hybrid storage system, manufactured by DLR in Stuttgart in 1993, had 200 MWht. It was designed for 4-h cycles (approximately 3-h charge and 1-h discharge). The heat transfer fluid used is a synthetic oil and PCM, sodium nitrate ($NaNO_3$, with a melting temperature of 310°C), concrete and a NaOH/NaCl mixture (with a melting temperature of 370°C) that have been used as storage materials. The most innovative aspects of this concept are the possibility of improving the efficiency of storage systems, because of a combination of good thermal conductivity, a reduced cost for concrete and promising PCM storage capacities and properties.

In order to increase energy storage efficiency, it is possible to adopt several PCM modules installed in cascade in order to operate over a wider range of temperature (cascade latent heat storage (CLHS)) (IEA ECES 2010; Crespo et al. 2019). Various numerical and experimental studies show the positive effect of CLHS technologies compared to non-cascaded systems because of the optimized use of various phase transition phenomena (Watanabe et al. 1993; Aceves et al. 1998; Michels and Pitz-Paal 2007; Liu et al. 2015; Prieto and Cabeza 2019).

When a LH-TESS CLHS is compared with a SH-TESS with two molten salts tanks, the following conclusions can be drawn (Michels and Pitz-Paal 2007):

– The LH-TESS CLHS provides the same storage capacity as the system with two tanks with a smaller quantity of molten salts.

– A heat exchanger is required with LH-TESS CLHS.

– LH-TESS CLHS provides a uniform output temperature of the heat transfer fluid.

– Storage systems with tanks require the addition of two pumps for the molten salts and also of a device for maintaining salts in liquid state, which is not necessary for CLHS technologies.

In conclusion, average-high temperature LH-TESS are being considered, though still in an exploratory stage, in the industry of CSP plants. Recent developments in the literature (Crespo et al. 2019) concern the study of new PCMs and various techniques for improving the thermal transfer in PCMs (such as PCM encapsulation, addition of nanoparticles, the use of metal foams, expanded graphite, heat pipes and metal fins as heat exchangers and even the design of heat exchangers in the temperature range between 120 and 400°C, etc.). According to these studies, it can be stated that LH-TESS technologies would allow a better storage capacity, PCMs having generally a higher heat storage density than the sensible heat materials. Consequently, LH-TESS require less space for storing the same quantity of energy as SH-TESS, which is a significant advantage in terms of systems compactness. Moreover, they are able to provide heat at nearly constant temperature, which is a requirement for the thermodynamic applications of CSP.

6.7. Other domains

Various application possibilities have already been considered in the automotive field (Alva et al. 2018; Nazir et al. 2019). Implementing LH-TES technologies may prove beneficial for many of them. First of all, user comfort is improved by maintaining the inside temperature at controlled values. According to the works of Jamekhorsid and Sadrameli (2012), adding a thin layer of PCM, integrated in the ceiling, allows a significant reduction of maximal temperature in the middle of summer. On a different note, Gumus (2009) developed an experimental LH-TES device integrating a hydrated salt as PCM. The purpose is to maintain the temperature of an internal combustion engine above 15°C during winter, which leads to a significant reduction of toxic emissions. Finally, with the emergence of electric vehicles, secondary energy consumption should be reduced, in order to increase vehicle autonomy. The study conducted by Wang et al. (2016) presents various designs of LH-TESS aimed at reducing air conditioning-related energy needs by adding PCMs. It shows that such a system integrated into electric vehicles could at least provide enough heat for maintaining a comfortable temperature in the cabin for a duration equivalent to the average daily ride of an American worker.

These developments are applicable to any other mode of transportation requiring air conditioning or meeting user comfort needs. It is particularly worth mentioning the progressive integration of these systems in the space sector (Collette et al. 2011; Lee et al. 2017).

Thermo-regulating textiles are another common application of LH-TESS (Alva et al. 2018; Nazir et al. 2019). The melting temperature of the PCMs included in these textiles generally ranges between 28 and 35°C. Their role is to absorb heat when the ambient temperature increases and release it when temperature decreases. Common applications of this type of technology include high altitude/extreme weather clothing and spacesuits, as well as thermo-regulating curtains and furniture.

There are three methods for incorporating PCMs into the textile material: in the form of fibers, by coating and by drafting (Alva et al. 2018):

– in the form of fibers, the micro-encapsulated PCM is added to a liquid solution of polymers which is then formed by various spinning techniques (dry spinning, wet spinning, electrospinning or extrusion);

– by coating, a textile substrate is coated with PCM. The PCM coating is prepared by mixing the micro-encapsulated PCM with a polymer binder, a surfactant, dispersing and anti-foaming agent and also a thickener;

– by drafting, the PCM is implemented in a film of polymers which is itself applied by drafting onto one side of the tissue.

Due to the large number of variables to be considered, it is difficult to evaluate the performances obtained for this type of application. While the heat retention effects are undeniable, as shown by the works of Shin et al. (2005), Babu and Arunraj (2018) or Chen et al. (2019), they are also accompanied by phenomena with variable effects (permeability, rigidity, deformation, cyclability) specific to this type of application.

In the field of electronics, cooling the components exposed to overheating is a critical issue (Alva et al. 2018; Nazir et al. 2019). Most of these components are currently used in combination with fin tube exchangers that dissipate heat by natural or forced convection and by radiation. The design of most coolers involves mobile and noisy parts, such as fans, which present many disadvantages in terms of performance, ergonomics or safety. According to Sahoo et al. (2016), the use of PCMs for cooling electronic components with intermittent operations was particularly adapted. These observations are confirmed by many studies that proved the usefulness of these technologies. Indeed, they significantly delay the occurrence of overheating and reduce the energy needs related to the use of mobile parts (Hosseinizadeh et al. 2011; Sahoo et al. 2016; Krishna et al. 2017; Hubert et al. 2021).

6.8. Conclusion

Overall, the development and integration of thermal energy storage systems in these applications are critical for the ongoing energy transition. Indeed, these systems make the connection between intermittent primary resources and user needs. However, there is a need to solve significant scientific and technological difficulties inherent to both source and usage: storage temperatures, thermal capacity, storage duration, energy and economic efficiencies, integration of systems into installations, etc.

There are three possibilities for thermal energy storage: by sensible heat, latent heat and thermochemical storage. Sensible heat storage systems are already widely available on the market, but have several drawbacks, most notably (a) low-energy density, particularly in storage applications over narrow ranges of temperature, leading to significant storage volumes and (b) heat losses generating difficulties in long-term thermal energy storage. As for thermochemical systems, they often have economic (high cost) and environmental (nonrenewable materials) drawbacks and require complex technologies.

The use of latent heat storage systems, integrating phase change materials, makes it possible to eliminate these drawbacks while providing essential advantages. Indeed, these technologies allow charge and discharge processes at quasi constant temperature, which is an important advantage in many applications. The energy densities of PCMs are higher than those of materials integrating sensible heat storage systems. The same amount of energy can be stored in a smaller volume, hence more compact systems are possible thanks to LH-TES. There is a wide diversity of PCMs, and therefore ranges of associated temperatures from negative temperatures to very high temperatures, which are applicable to many domains. In most cases, the associated systems are relatively simple technologies (except at high temperatures).

Annex 30 of IEA (IEA ECES 2018) lists the various storage systems in terms of prototypes, demonstrators and applications. This report compares the maturity of technologies to assess their transferability to the market for applications in the heat networks, commercial (non-residential) areas, solar power plants and industrial applications.

Table 6.2 presents a non-exhaustive list of heat storage systems, their degree of maturity and also their main technical, environmental and safety drawbacks (Megret et al. 2011).

Among the various types of latent heat storage systems, those allowing cold storage integrate water with or without additives as the main storage material. These technologies have already been commercialized and tested, particularly for applications in buildings and heat networks and also for industrial applications (agrifood industry, etc.). These systems have significant advantages related to the nature of the storage material (available, cheap, safe, renewable, etc.) and most often use simple technologies. However, as in the case of sensible heat storage systems, long-term energy storage is difficult due to significant heat losses.

Sectors	Habitat, tertiary 90°C			IAA, bleachwork 200°C		Metallurgy, CSP 350°C	
Storage Techniques	Bed packed with PCM paraffin	Bed packed with other PCMs (hydrated salts, eutectics)	Storage with graphite/ paraffin matrix	Bed packed with HT PCM (eutectics, salts)	Storage with graphite/salts matrix	Storage in tank with HT PCM (eutectics, salts)	Cascade LHTS
Storage Temperature	T < 100°C	50–400°C (depending on transfer fluid: water, oil and PCM type)	T < 100°C	110°C < T < 400°C	110°C < T < 400°C	110°C < T < 400°C	300°C < T < 400°C
Storage Density (kWh/m^3)	From 40 to 80	From 50 to 80	From 40 to 80	From 50 to 80	From 60 to 150	From 60 to 150	–
Thermal Conductivity (W/m/K)	About 0.2	About 0.1	From 1 to 10 depending on the PCM/ graphite percentage	0.1	From 1 to 15 depending on the PCM/graphite percentage	0.1	–

Sectors	Habitat, tertiary 90°C		IAA, bleachwork 200°C		Metallurgy, CSP 350°C	
Technological, environmental and security drawbacks	Recyclability, lifetime, origin (petroleum derivatives), low thermal conductivity of paraffin	Recyclability, lifetime, low thermal conductivity	Recyclability, lifetime, origin (petroleum derivatives), low thermal conductivity of paraffin	ICPE regulation, corrosion, low thermal conductivity	–	ICPE regulation, corrosion, low thermal conductivity	–
Cost	High	High	High	–	High	–	High
Development level	Materials on the market but relatively new technology	Technology and materials on the market but relatively new technology	Prototype (DLR) – Materials on the market	Technology and materials on the market but relatively new technology	Prototype (DLR)	Technology and materials on the market but relatively new technology	Research prototype

Table 6.2. *Characteristics of latent heat storage technologies depending on sector (adapted from Megret et al. (2011))*

Passive and active latent heat storage systems have also been the subject of extensive research aiming at their integration in various sectors (building, transportation, textiles, electronics and more recently in nuclear or cogeneration power plants). These research efforts are still ongoing focusing on major problems that have not yet been solved mainly due to materials integrated in latent heat storage systems. Indeed, they may have drawbacks of various types: economic (high cost), environmental (nonrenewable materials) and technical (low thermal conductivity, etc.). Security problems (flammability) or corrosion may also occur.

Finally, research efforts have started for the integration of latent heat storage systems in CSP plants. Their aim is to combine or replace the existing sensible heat storage systems with latent heat systems. The challenge to be addressed is great due to the range of temperatures (high temperatures) specific to this type of application.

Latent heat energy storage systems have a strong potential. They may in fact be highly competitive once the above-mentioned current problems are solved.

6.9. References

Aceves, S.M., Nakamura, H., Reistad, G.M., Martinez-Frias, J. (1998). Optimization of a class of latent thermal energy storage systems with multiple phase-change materials. *J. Sol. Energy Eng.*, 120, 14–19.

ADEME (2011). Feuille de route solaire thermodynamique.

Adesina, A. (2019). Use of phase change materials in concrete: Current challenges. *Renew. Energy Environ. Sustain.*, 4, 9.

Akeiber, H., Nejat, P., Majid, M.Z.A., Wahid, M.A., Jomehzadeh, F., Zeynali Famileh, I., Calautit, J.K., Hughes, B.R., Zaki, S.A. (2016). A review on phase change material (PCM) for sustainable passive cooling in building envelopes. *Renewable and Sustainable Energy Reviews*, 60, 1470–1497.

Alameri, S.A. (2015). A coupled nuclear reactor thermal energy storage system for enhanced load following operation. Thesis, Colorado School of Mines, Golden.

Alehosseini, E. and Jafari, S.M. (2019). Micro/nano-encapsulated phase change materials (PCMs) as emerging materials for the food industry. *Trends in Food Science & Technology*, 91, 116–128.

Alva, G., Lin, Y., Fang, G. (2018). An overview of thermal energy storage systems. *Energy*, 144, 341–378.

Arivazhagan, R., Arun Prakash, S., Kumaran, P., Sankar, S., Loganathan, G.B., Arivarasan, A. (2019). Performance analysis of concrete block integrated with PCM for thermal management. In *Materials Today: Proceedings*, 22(3), 370–374.

Babu, V.R. and Arunraj, A. (2018). Thermo regulated clothing with phase change materials. *Journal of Textile Engineering & Fashion Technology*, 4.

Barnes, F.S. and Levine, J.G. (2017). *Large Energy Storage Systems Handbook*. CRC Press, Boca Raton.

Benard, C., Body, Y., Gobin, D., Guerrier, B. (1982). Use of a variable parameter test-cell for the study of latent-heat solar walls. *Solar Energy*, 29, 101–109.

Berardi, U. and Gallardo, A.A. (2019). Properties of concretes enhanced with phase change materials for building applications. *Energy and Buildings*, 199, 402–414.

Bouhal, T., El Rhafiki, T., Kousksou, T., Jamil, A., Zeraouli, Y. (2018). PCM addition inside solar water heaters: Numerical comparative approach. *Journal of Energy Storage*, 19, 232–246.

Bourdeau, L. (1982). Utilisation d'un matériau à changement de phase dans un mur Trombe sans thermocirculation. *Revue de physique appliquée*, 17, 633–642.

Bourdillon, A.C., Verdin, P.G., Thompson, C.P. (2015). Numerical simulations of water freezing processes in cavities and cylindrical enclosures. *Applied Thermal Engineering*, 75, 839–855.

Brückner, S., Liu, S., Miró, L., Radspieler, M., Cabeza, L.F., Lävemann, E. (2015). Industrial waste heat recovery technologies: An economic analysis of heat transformation technologies. *Applied Energy*, 151, 157–167.

Cabeza, L.F., Ibáñez, M., Solé, C., Roca, J., Nogués, M. (2006). Experimentation with a water tank including a PCM module. *Solar Energy Materials and Solar Cells*, 90, 1273–1282.

Castellón, C., Martorell, I., Cabeza, L.F., Fernández, A.I., Manich, A.M. (2011). Compatibility of plastic with phase change materials (PCM). *International Journal of Energy Research*, 35, 765–771.

Chaiyat, N. (2015). Energy and economic analysis of a building air-conditioner with a phase change material (PCM). *Energy Conversion and Management*, 94, 150–158.

Charvát, P., Klimeš, L., Zálešák, M. (2019). Utilization of an Air-PCM heat exchanger in passive cooling of buildings: A simulation study on the energy saving potential in different European climates. *Energies*, 12, 1133.

Chen, S.-L., Chen, C.-L., Tin, C.-C., Lee, T.-S., Ke, M.-C. (2000). An experimental investigation of cold storage in an encapsulated thermal storage tank. *Experimental Thermal and Fluid Science*, 23, 133–144.

Chen, X., Wu, B., Fan, Y., Duan, X., Yang, M. (2019). Temperature adjusting performance of thermoregulated woven fabric finished with phase-change microcapsule in low-temperature environment. *Journal of Engineered Fibers and Fabrics*.

Coleman, J., Bragg-Sitton, S., Dufek, E., Johnson, S., Rhodes, J., Davidson, T., Webber, M.E. (2017). An evaluation of energy storage options for nuclear power. Technical Report No. INL/EXT-17-42420, Idaho National Laboratory, Idaho Falls.

Collette, J.-P., Rochus, P., Peyrou-Lauga, M.R., Crahay, M.J. (2011). Phase change material device for spacecraft thermal control. In *62nd International Astronautical Congress*, Cape Town.

Crespo, A., Barreneche, C., Ibarra, M., Platzer, W. (2019). Latent thermal energy storage for solar process heat applications at medium-high temperatures – A review. *Solar Energy, Thermal Energy Storage for Solar Applications*, 192, 3–34.

D'Alessandro, A., Pisello, A.L., Fabiani, C., Ubertini, F., Cabeza, L.F., Cotana, F. (2018). Multifunctional smart concretes with novel phase change materials: Mechanical and thermo-energy investigation. *Applied Energy*, 212, 1448–1461.

Dardir, M., El Mankibi, M., Haghighat, F., Klimes, L. (2019a). Development of PCM-to-air heat exchanger for integration in building envelope-modeling and validation. *Solar Energy*, 190, 367–385.

Dardir, M., Panchabikesan, K., Haghighat, F., El Mankibi, M., Yuan, Y. (2019b). Opportunities and challenges of PCM-to-air heat exchangers (PAHXs) for building free cooling applications – A comprehensive review. *Journal of Energy Storage*, 22, 157–175.

Delgado, J.M.P.Q., Martinho, J.C., Vaz Sá, A., Guimarães, A.S., Abrantes, V. (2019). PCM current applications and thermal performance. In *Thermal Energy Storage with Phase Change Materials: A Literature Review of Applications for Buildings Materials*, Delgado, J.M.P.Q., Martinho, J.C., Vaz Sá, A., Guimarães, A.S., Abrantes, V. (eds). Springer, Cham.

Devahastin, S. and Pitaksuriyarat, S. (2006). Use of latent heat storage to conserve energy during drying and its effect on drying kinetics of a food product. *Applied Thermal Engineering*, 26, 1705–1713.

Faraj, K., Khaled, M., Faraj, J., Hachem, F., Castelain, C. (2019). Phase change material thermal energy storage systems for cooling applications in buildings: A review. *Renewable and Sustainable Energy Reviews*, 109579.

Faraj, K., Khaled, M., Faraj, J., Hachem, F., Castelain, C. (2021). A review on phase change materials for thermal energy storage in buildings: Heating and hybrid applications. *Journal of Energy Storage*, 33, 101913.

Farouk, B. and Guceri, S.I. (1981). Trombe-Michel wall using phase change materials. In *Proceedings of the Second Miami International Conference*, Miami Beach.

Fazilati, M.A. and Alemrajabi, A.A. (2013). Phase change material for enhancing solar water heater, an experimental approach. *Energy Conversion and Management*, 71, 138–145.

Ferrière, A. (2008). Centrales solaires thermodynamiques. *Techniques de l'ingénieur*. Document No. TIB594DUO.

Fiorito, F. (2012). Trombe walls for lightweight buildings in temperate and hot climates. Exploring the use of phase-change materials for performances improvement. In *Energy Procedia, 1st International Conference on Solar Heating and Cooling for Buildings and Industry (SHC 2012)*.

Flores, J.F.C., Espagnet, A.R., Chiu, J.N., Martin, V., Lacarrière, B. (2017). Techno-economic assessment of active latent heat thermal energy storage systems with low-temperature district heating. *International Journal of Sustainable Energy Planning and Management*, 13, 5–18.

Forster, M. (2004). Theoretical investigation of the system SnOx/Sn for the thermochemical storage of solar energy. *Energy, SolarPACES*, 29, 789–799.

Gao, H., Wang, J., Chen, X., Wang, G., Huang, X., Li, A., Dong, W. (2018). Nanocon-finement effects on thermal properties of nanoporous shape-stabilized composite PCMs: A review. *Nano Energy*, 53, 769–797.

Geetha, N.B. and Velraj, R. (2012). Passive cooling methods for energy efficient buildings with and without thermal energy storage – A review. *Energy Education Science and Technology Part A: Energy Science and Research*, 29, 913–946.

Gibb, D., Johnson, M., Romaní, J., Gasia, J., Cabeza, L.F., Seitz, A. (2018). Process integration of thermal energy storage systems – Evaluation methodology and case studies. *Applied Energy*, 230, 750–760.

Goldstern, W. (1979). Thermal energy storage in industry and power stations. *Journal of the Institute of Energy*, 52, 185–192.

Gomez, J.C. (2011). High-temperature phase change materials (PCM) candidates for thermal energy storage (TES) applications (No. NREL/TP-5500-51446, 1024524).

Gumus, M. (2009). Reducing cold-start emission from internal combustion engines by means of thermal energy storage system. *Applied Thermal Engineering*, 29, 652–660.

Hamdan, M.A. and Elwerr, F.A. (1996). Thermal energy storage using a phase change material. *Solar Energy*, 56, 183–189.

Hawes, D.W., Feldman, D., Banu, D. (1993). Latent heat storage in building materials. *Energy and Buildings*, 20, 77–86.

Hawlader, M.N.A., Uddin, M.S., Khin, M.M. (2003). Microencapsulated PCM thermal-energy storage system. *Applied Energy, Energex 2002 – New and Renewable Sources of Energy – Topic I*, 74, 195–202.

Hed, G. and Bellander, R. (2006). Mathematical modelling of PCM air heat exchanger. *Energy and Buildings*, 38, 82–89.

Heinen, S., Mancarella, P., O'Dwyer, C., O'Malley, M. (2018). Heat electrification: The latest research in Europe. *IEEE Power and Energy Magazine*, 16, 69–78.

Herrmann, U., Kelly, B., Price, H. (2004). Two-tank molten salt storage for parabolic trough solar power plants. *Energy*, 29, 883–893.

Hoffmann, J.-F. (2015). Stockage thermique pour centrale solaire thermodynamique à concentration mettant en œuvre des matériaux céramiques naturels ou recyclés. PhD Thesis, Université de Perpignan, Perpignan.

Hoffmann, J.-F., Henry, J.-F., Vaitilingom, G., Olives, R., Chirtoc, M., Caron, D., Py, X. (2016). Temperature dependence of thermal conductivity of vegetable oils for use in concentrated solar power plants, measured by 3omega hot wire method. *International Journal of Thermal Sciences*, 107, 105–110.

Hosseinizadeh, S.F., Tan, F.L., Moosania, S.M. (2011). Experimental and numerical studies on performance of PCM-based heat sink with different configurations of internal fins. *Applied Thermal Engineering*, 31, 3827–3838.

Hubert, R., Bou Matar, O., Foncin, J., Coquet, P., Tan, D., Li, H., Teo, E.H.T., Merlet, T., Pernod, P. (2021). An effective thermal conductivity model for architected phase change material enhancer: Theoretical and experimental investigations. *International Journal of Heat and Mass Transfer*, 176, 121364.

Hutty, T.D., Patel, N., Dong, S., Brown, S. (2020). Can thermal storage assist with the electrification of heat through peak shaving? In *4th Annual CDT Conference in Energy Storage & Its Applications*.

IEA ECES (2010). Optimised industrial process heat and power generation with thermal energy storage. Final report.

IEA ECES (2018). Annex 30 Thermal energy storage for cost-effective energy management. Report.

Iten, M. (2021). Techno-economic assessment of an air-multiple PCM active storage unit for free cooling application. *Sustainability*, 13, 12936.

Jacob, R. and Bruno, F. (2015). Review on shell materials used in the encapsulation of phase change materials for high temperature thermal energy storage. *Renewable and Sustainable Energy Reviews*, 48, 79–87.

Jamekhorshid, A. and Sadrameli, S.M. (2012). Application of phase change materials (PCMs) in maintaining comfort temperature inside an automobile. *International Journal of Chemical and Molecular Engineering*, 6, 33–35.

Javier Batlles, F., Puertas, A.M., Romero-Cano, M.S., Rosiek, S., Gil, B., Kasperski, J., Nemś, A., Nemś, M., Grágeda, M., Ushak, S. et al. (2021). Application of thermal storage in over-night refrigeration of an institutional building. *Solar Energy*, 220, 450–461.

Johnson, M., Vogel, J., Hempel, M., Dengel, A., Seitz, M., Hachmann, B. (2015). High temperature latent heat thermal energy storage integration in a Co-gen plant. In *Energy Procedia, 9th International Renewable Energy Storage Conference, IRES*.

Johnson, M., Vogel, J., Hempel, M., Hachmann, B., Dengel, A. (2017). Design of high temperature thermal energy storage for high power levels. *Sustainable Cities and Society*, 35, 758–763.

Kaasinen, H. (1992). The absorption of phase change substances into commonly used building materials. *Solar Energy Materials and Solar Cells, Special Issue on Heat Storage Materials*, 27, 173–179.

Kavvadias, K., Jimenez Navarro, J.P., Zucker, A. (2018). Case study on the impact of cogeneration and thermal storage on the flexibility of the power system (No. EUR 29082 EN). Publications Office of the European Union.

Kearney, D., Herrmann, U., Nava, P., Kelly, B., Mahoney, R., Pacheco, J., Cable, R., Potrovitza, N., Blake, D., Price, H. (2003). Assessment of a molten salt heat transfer fluid in a parabolic trough solar field. *J. Sol. Energy Eng.*, 125, 170–176.

Kearney, D., Kelly, B., Herrmann, U., Cable, R., Pacheco, J., Mahoney, R., Price, H., Blake, D., Nava, P., Potrovitza, N. (2004). Engineering aspects of a molten salt heat transfer fluid in a trough solar field. *Energy, SolarPACES*, 29, 861–870.

Kılıçkap, S., El, E., Yıldız, C. (2018). Investigation of the effect on the efficiency of phase change material placed in solar collector tank. *Thermal Science and Engineering Progress*, 5, 25–31.

Knebel, D.E. (1995). Predicting and evaluating the performance of ice-harvesting thermal energy storage systems. In *ASHRAE Transactions.*

Kosny, J. (2015a). *PCM-Enhanced Building Components: An Application of Phase Change Materials in Building Envelopes and Internal Structures, Engineering Materials and Processes*. Springer, Cham.

Kosny, J. (2015b). Short history of PCM applications in building envelopes. In *PCM-Enhanced Building Components: An Application of Phase Change Materials in Building Envelopes and Internal Structures, Engineering Materials and Processes*, Kośny, J. (ed.). Springer, Cham.

Krishna, J., Kishore, P.S., Solomon, A.B. (2017). Heat pipe with nano enhanced-PCM for electronic cooling application. *Experimental Thermal and Fluid Science*, 81, 84–92.

Kurata, O., Iki, N., Matsunuma, T., Maeda, T., Hirano, S., Kadoguchi, K., Takeuchi, H., Yoshida, H. (2014a). Micro gas turbine cogeneration system with latent heat storage at the University: Part I: Plan and energy flow test. *Applied Thermal Engineering*, 65, 513–523.

Kurata, O., Iki, N., Matsunuma, T., Maeda, T., Hirano, S., Kadoguchi, K., Takeuchi, H., Yoshida, H. (2014b). Micro gas turbine cogeneration system with latent heat storage at the University: Part II: Part load and thermal priority mode. *Applied Thermal Engineering*, 65, 246–254.

Kurata, O., Iki, N., Matsunuma, T., Maeda, T., Hirano, S., Kadoguchi, K., Takeuchi, H., Yoshida, H. (2014c). Micro gas turbine cogeneration system with latent heat storage at the University: Part III: Temperature control schedule. *Applied Thermal Engineering*, 70, 705–715.

Kuravi, S., Trahan, J., Goswami, D.Y., Rahman, M.M., Stefanakos, E.K. (2013). Thermal energy storage technologies and systems for concentrating solar power plants. *Progress in Energy and Combustion Science*, 39, 285–319.

Kuznik, F., David, D., Johannes, K., Roux, J.-J. (2011). A review on phase change materials integrated in building walls. *Renewable and Sustainable Energy Reviews*, 15, 379–391.

Laing, D., Steinmann, W.-D., Tamme, R., Richter, C. (2006). Solid media thermal storage for parabolic trough power plants. *Solar Energy, Solar Power and Chemical Energy Systems (SolarPACES'04)*, 80, 1283–1289.

Lee, K.-L., Tarau, C., Velson, N.V. (2017). Development of a heat exchanger with integrated thermal storage for spacecraft thermal management applications. In *47th International Conference on Environmental Systems*, Charleston.

Liu, M., Saman, W., Bruno, F. (2012). Review on storage materials and thermal performance enhancement techniques for high temperature phase change thermal storage systems. *Renewable and Sustainable Energy Reviews*, 16, 2118–2132.

Liu, M., Tay, N.H.S., Belusko, M., Bruno, F. (2015). Investigation of cascaded shell and tube latent heat storage systems for solar tower power plants. *Energy Procedia*, 69, 913–924.

Liu, Z., Yu, Z., Yang, T., Qin, D., Li, S., Zhang, G., Haghighat, F., Joybari, M.M. (2018). A review on macro-encapsulated phase change material for building envelope applications. *Building and Environment*, 144, 281–294.

Lu, Y.L., Zhang, W.H., Yuan, P., Xue, M.D., Qu, Z.G., Tao, W.Q. (2010). Experimental study of heat transfer intensification by using a novel combined shelf in food refrigerated display cabinets (experimental study of a novel cabinets). *Applied Thermal Engineering*, 30, 85–91.

Madad, A., Mouhib, T., Mouhsen, A. (2018). Phase change materials for building applications: A thorough review and new perspectives. *Buildings*, 8, 63.

Matsumoto, K., Namiki, Y., Okada, M., Kawagoe, T., Nakagawa, S., Kang, C. (2004). Continuous ice slurry formation using a functional fluid for ice storage. *International Journal of Refrigeration*, 27, 73–81.

Mazman, M., Cabeza, L.F., Mehling, H., Nogues, M., Evliya, H., Paksoy, H.Ö. (2009). Utilization of phase change materials in solar domestic hot water systems. *Renewable Energy*, 34, 1639–1643.

Mazzoni, S., Sze, J.Y., Nastasi, B., Ooi, S., Desideri, U., Romagnoli, A. (2021). A techno-economic assessment on the adoption of latent heat thermal energy storage systems for district cooling optimal dispatch & operations. *Applied Energy*, 289, 116646.

Medrano, M., Gil, A., Martorell, I., Potau, X., Cabeza, L.F. (2010). State of the art on high-temperature thermal energy storage for power generation. Part 2 – Case studies. *Renewable and Sustainable Energy Reviews*, 14, 56–72.

Megret, O., Bequet, L., Manificat, A., Weber, C. (2011). Transport longue distance et stockage d'énergie calorifique, en couplage avec des procédés de valorisation énergétique des déchets – État de l'art (No. 09-0233/1A). Record.

Mehling, H. and Cabeza, L.F. (2008). *Heat and Cold Storage with PCM: An Up to Date Introduction into Basics and Applications*. Springer, Berlin.

Mehling, H., Cabeza, L.F., Hippeli, S., Hiebler, S. (2003). PCM-module to improve hot water heat stores with stratification. *Renewable Energy*, 28, 699–711.

Memarian, S., Kari, B.M., Fayaz, R., Asadi, S. (2018). Single and combined phase change materials: Their effect on seasonal transition period. *Energy and Buildings*, 169, 453–472.

Memon, S.A. (2014). Phase change materials integrated in building walls: A state of the art review. *Renewable and Sustainable Energy Reviews*, 31, 870–906.

Michels, H. and Pitz-Paal, R. (2007). Cascaded latent heat storage for parabolic trough solar power plants. *Solar Energy*, 81, 829–837.

Miró, L., Brückner, S., Cabeza, L.F. (2015). Mapping and discussing Industrial Waste Heat (IWH) potentials for different countries. *Renewable and Sustainable Energy Reviews*, 51, 847–855.

Mourid, A. and El Alami, M. (2017). Thermal behavior of a building provided with phase-change materials on the roof and exposed to solar radiation. *J. Sol. Energy Eng.*, 139.

Najafian, A., Haghighat, F., Moreau, A. (2015). Integration of PCM in domestic hot water tanks: Optimization for shifting peak demand. *Energy and Buildings, SI: IEA ECES Annex 31 Special Issue on Thermal Energy Storage*, 106, 59–64.

Nazir, H., Batool, M., Bolivar Osorio, F.J., Isaza-Ruiz, M., Xu, X., Vignarooban, K., Phelan, P., Kannan, A.M. (2019). Recent developments in phase change materials for energy storage applications: A review. *International Journal of Heat and Mass Transfer*, 129, 491–523.

Niyas, H. and Muthukumar, P. (2018). A novel heat transfer enhancement technique for performance improvements in encapsulated latent heat storage system. *Solar Energy*, 164, 276–286.

Nkwetta, D.N., Vouillamoz, P.-E., Haghighat, F., El Mankibi, M., Moreau, A., Desai, K. (2014). Phase change materials in hot water tank for shifting peak power demand. *Solar Energy*, 107, 628–635.

Omara, A.A.M. and Abuelnuo, A.A.A. (2019). Trombe walls with phase change materials: A review. *Energy Storage*, e123.

Ortega Del Rosario, M.D.L.A., Nadeau, J.-P., Bruneau, D., Sebastian, P., Sommier, A., Lopez, J. (2016). Design and optimization of PCM-air heat exchangers. In *Virtual Concept Workshop 2016: Major Trends in Product Design*.

Pacheco, J.E., Showalter, S.K., Kolb, W.J. (2002). Development of a molten-salt thermocline thermal storage system for parabolic trough plants. *J. Sol. Energy Eng.*, 124, 153–159.

Pelay, U., Luo, L., Fan, Y., Stitou, D., Rood, M. (2017). Thermal energy storage systems for concentrated solar power plants. *Renewable and Sustainable Energy Reviews*, 79, 82–100.

Pop, O.G., Fechete Tutunaru, L., Bode, F., Abrudan, A.C., Balan, M.C. (2018). Energy efficiency of PCM integrated in fresh air cooling systems in different climatic conditions. *Applied Energy*, 212, 976–996.

Prieto, C. and Cabeza, L.F. (2019). Thermal energy storage (TES) with phase change materials (PCM) in solar power plants (CSP). Concept and plant performance. *Applied Energy*, 254, 113646.

Rastogi, M., Chauhan, A., Vaish, R., Kishan, A. (2015). Selection and performance assessment of phase change materials for heating, ventilation and air-conditioning applications. *Energy Conversion and Management*, 89, 260–269.

Raul, A., Saha, S.K., Jain, M. (2020). Transient performance analysis of concentrating solar thermal power plant with finned latent heat thermal energy storage. *Renewable Energy*, 145, 1957–1971.

Rismanchi, B., Saidur, R., BoroumandJazi, G., Ahmed, S. (2012). Energy, exergy and environmental analysis of cold thermal energy storage (CTES) systems. *Renewable and Sustainable Energy Reviews*, 16, 5741–5746.

Rouault, F., Bruneau, D., Sebastian, P., Lopez, J. (2013). Numerical modelling of tube bundle thermal energy storage for free-cooling of buildings. *Applied Energy*, 111, 1099–1106.

Rouault, F., Bruneau, D., Sebastian, P., Nadeau, J.-P. (2016). Use of a latent heat thermal energy storage system for cooling a light-weight building: Experimentation and co-simulation. *Energy and Buildings*, 127, 479–487.

Sabihuddin, S., Kiprakis, A.E., Mueller, M. (2015). A numerical and graphical review of energy storage technologies. *Energies*, 8, 172–216.

Sahoo, S.K., Das, M.K., Rath, P. (2016). Application of TCE-PCM based heat sinks for cooling of electronic components: A review. *Renewable and Sustainable Energy Reviews*, 59, 550–582.

Said, M.A. and Hassan, H. (2021). Impact of energy storage of new hybrid system of phase change materials combined with air-conditioner on its heating and cooling performance. *Journal of Energy Storage*, 36, 102400.

Sardeshpande, V. and Pillai, I.R. (2012). Effect of micro-level and macro-level factors on adoption potential of solar concentrators for medium temperature thermal applications. *Energy for Sustainable Development*, 16, 216–223.

Sawadogo, M., Duquesne, M., Belarbi, R., Hamami, A.E.A., Godin, A. (2021). Review on the integration of phase change materials in building envelopes for passive latent heat storage. *Applied Sciences*, 11, 9305.

Sayyar, M., Weerasiri, R.R., Soroushian, P., Lu, J. (2014). Experimental and numerical study of shape-stable phase-change nanocomposite toward energy-efficient building constructions. *Energy and Buildings*, 75, 249–255.

Seshie, Y.M., N'Tsoukpoe, K.E., Neveu, P., Coulibaly, Y., Azoumah, Y.K. (2018). Small scale concentrating solar plants for rural electrification. *Renewable and Sustainable Energy Reviews*, 90, 195–209.

Shafigh, P., Asadi, I., Mahyuddin, N.B. (2018). Concrete as a thermal mass material for building applications – A review. *Journal of Building Engineering*, 19, 14–25.

Sharma, R.K., Ganesan, P., Tyagi, V.V., Metselaar, H.S.C., Sandaran, S.C. (2015). Developments in organic solid-liquid phase change materials and their applications in thermal energy storage. *Energy Conversion and Management*, 95, 193–228.

Shin, Y., Yoo, D.-I., Son, K. (2005). Development of thermoregulating textile materials with microencapsulated phase change materials (PCM). IV. Performance properties and hand of fabrics treated with PCM microcapsules. *Journal of Applied Polymer Science*, 97, 910–915.

Shirinbakhsh, M., Mirkhani, N., Sajadi, B. (2018). A comprehensive study on the effect of hot water demand and PCM integration on the performance of SDHW system. *Solar Energy*, 159, 405–414.

Shnaider, D.A., Divnich, P.N., Vakhromeev, I.E. (2010). Modeling the dynamic mode of steam accumulator. *Autom Remote Control*, 71, 1994–1998.

Soares, N., Costa, J.J., Gaspar, A.R., Santos, P. (2013). Review of passive PCM latent heat thermal energy storage systems towards buildings' energy efficiency. *Energy and Buildings*, 59, 82–103.

Steinmann, W.-D. and Eck, M. (2006). Buffer storage for direct steam generation. *Solar Energy, Solar Power and Chemical Energy Systems (SolarPACES'04)*, 80, 1277–1282.

Steinmann, W.-D. and Tamme, R. (2008). Latent heat storage for solar steam systems. *J. Sol. Energy Eng.*, 130.

Stevanovic, V.D., Maslovaric, B., Prica, S. (2012). Dynamics of steam accumulation. *Applied Thermal Engineering*, 37, 73–79.

Talmatsky, E. and Kribus, A. (2008). PCM storage for solar DHW: An unfulfilled promise? *Solar Energy*, 82, 861–869.

Tamme, R., Taut, U., Streuber, C., Kalfa, H. (1991). Energy storage development for solar thermal processes. *Solar Energy Materials*, 24, 386–396.

Torregrosa-Jaime, B., López-Navarro, A., Corberán, J.M., Esteban-Matías, J.C., Klinkner, L., Payá, J. (2013). Experimental analysis of a paraffin-based cold storage tank. *International Journal of Refrigeration*, 36, 1632–1640.

Trombe, F. (1974). Maisons solaires. *Techniques de l'ingénieur*, 3.

Tyagi, V.V. and Buddhi, D. (2007). PCM thermal storage in buildings: A state of art. *Renewable and Sustainable Energy Reviews*, 11, 1146–1166.

Wang, M. and Kusumoto, N. (2001). Ice slurry based thermal storage in multifunctional buildings. *Heat and Mass Transfer*, 37, 597–604.

Wang, X., Zhang, Y., Xiao, W., Zeng, R., Zhang, Q., Di, H. (2009). Review on thermal performance of phase change energy storage building envelope. *Chin. Sci. Bull.*, 54, 920–928.

Wang, M., Wolfe, E., Craig, T., Laclair, T.J., Abdelaziz, O., Gao, Z. (2016). Design and testing of a thermal storage system for electric vehicle cabin heating. In *SAE 2016 World Congress and Exhibition*.

Watanabe, T., Kikuchi, H., Kanzawa, A. (1993). Enhancement of charging and discharging rates in a latent heat storage system by use of PCM with different melting temperatures. *Heat Recovery Systems and CHP*, 13, 57–66.

Wijeysundera, N.E., Hawlader, M.N.A., Andy, C.W.B., Hossain, M.K. (2004). Ice-slurry production using direct contact heat transfer. *International Journal of Refrigeration*, 27, 511–519.

Xie, L., Tian, L., Yang, L., Lv, Y., Li, Q. (2017). Review on application of phase change material in water tanks. *Advances in Mechanical Engineering*.

Xu, H., Lin, W.Y., Dal Magro, F., Li, T., Py, X., Romagnoli, A. (2019). Towards higher energy efficiency in future waste-to-energy plants with novel latent heat storage-based thermal buffer system. *Renewable and Sustainable Energy Reviews*, 112, 324–337.

Yamagishi, Y., Takeuchi, H., Pyatenko, A.T., Kayukawa, N. (1999). Characteristics of microencapsulated PCM slurry as a heat-transfer fluid. *AIChE Journal*, 45, 696–707.

Zeinelabdein, R., Omer, S., Gan, G. (2018). Critical review of latent heat storage systems for free cooling in buildings. *Renewable and Sustainable Energy Reviews*, 82, 2843–2868.

Zeinelabdein, R., Omer, S., Gan, G. (2019). Experimental performance of latent thermal energy storage for sustainable cooling of buildings in hot-arid regions. *Energy and Buildings*, 186, 169–185.

7

Use of Hydrates for Cold Storage and Distribution in Refrigeration and Air-Conditioning Applications

Anthony DELAHAYE and Laurence FOURNAISON
INRAE, FRISE, Université Paris-Saclay, Antony, France

7.1. Introduction

Accounting for nearly 20% of electricity consumed worldwide (IIF-IIR 2019), cold production is one of the most energy intensive sectors. Additionally, 90% of refrigeration systems continue to use large amounts of greenhouse refrigerants (such as hydrofluorocarbons (HFCs)) and are thus heavily regulated (European Parliament, Council of the European Union 2015).

Nevertheless, the use of refrigeration for thermal control and process safety is still a basic need for development in many sectors (food, air conditioning, electronics, etc.). The refrigeration industry must therefore face both major environmental and economic challenges, and for this purpose must rely on technological innovation.

Among the solutions to reduce refrigeration carbon footprint, it is possible to limit harmful primary refrigerants by using environment-friendly fluids in secondary refrigeration.

Heat and Cold Storage 1,
coordinated by Pierre ODRU, Elena PALOMO DEL BARRIO.

Nevertheless, as a process of indirect cooling, secondary refrigeration requires additional components (heat exchanger, circulation pumps, etc.) that generate exergy losses. In order to improve secondary system efficiency, it is possible to use solid–liquid secondary refrigerants, such as hydrate slurries. These fluids are suspensions of phase change material (PCM) particles dispersed in a carrying liquid phase. The term PCM means that the particles melt or crystallize in a given temperature range, which allows them to absorb or release large amounts of thermal energy by latent heat.

The main advantage of slurries over conventional monophasic fluids is therefore their ability to carry much larger amounts of cold. Thus, for a given cooling capacity, pipe diameters or fluid flow rates can be reduced, which limits circulation-pump power consumption as well as exergy losses. Furthermore, the thermal stability of the slurry, due to the phase change, during a transport cycle limits thermodynamic irreversibilities in the evaporator of the refrigerating machine and ensures a better thermal comfort for the user.

Slurries do not only carry the cold, but they can also be used as thermal energy storage material in storage devices, which improve the flexibility, design and energy efficiency of refrigeration processes. Most industrial plants thus comprise a slurry generator and a distribution network associated with storage tanks containing the slurry.

The implementation of slurries, however, requires accurate online monitoring of the solid concentration, an essential parameter for the optimal management of the energy content of the fluid.

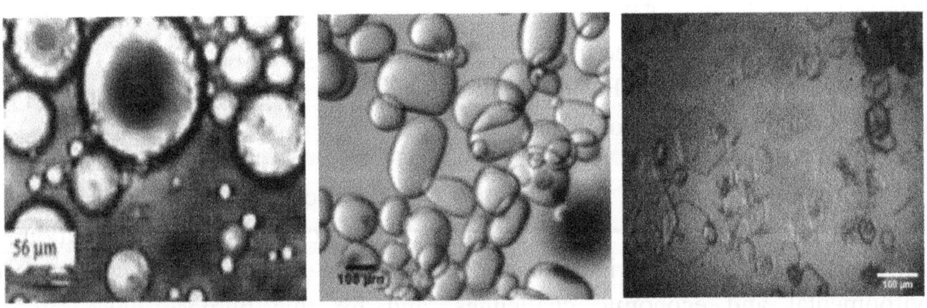

a) b) c)

Figure 7.1. *Illustration on a microscopic scale of different kinds of slurry: (a) paraffin slurry (Xu et al. 2005); (B) ice slurry (Stamatiou et al. 2005); (C) hydrate slurry (Clain et al. 2015)*

In the field of refrigeration and air conditioning, as illustrated by Figure 7.1, the solid–liquid secondary refrigerants can take many forms depending on the type of PCM used (Zhang et al. 2010b; Delgado et al. 2012; Li et al. 2012; Youssef et al. 2013):

– Paraffin slurry: it is made of fine paraffin particles dispersed in an aqueous solution, stabilized with surfactants (anionic or non-ionic).

– Microencapsulated slurry: it is made of PCM (paraffin) dispersed in an inert liquid, each droplet being individualized by a coating.

– Stabilized particle slurry: it consists of millimeter organic gel beads containing a PCM (paraffin + polymer) dispersed in water.

– Ice slurry: it is formed of ice crystals dispersed in an alcohol or salt-based aqueous solution at temperatures below 273 K. Ice slurry has a higher energy density than paraffin slurry.

– Hydrate slurry: it is made up of hydrate (clathrate, semiclathrate) crystals dispersed in a carrying liquid, stable on a wide range of temperatures, including temperatures above 273 K, and exhibiting an energy density comparable to or higher than that of ice slurries (Chatti et al. 2005). Other materials, also called salt hydrates, are used for refrigeration. These salt hydrates, which do not have a clathrate or semiclathrate structure, are not covered in this chapter.

In this chapter, the definition and the structural and thermodynamic properties of hydrates are first presented. Then, a review of hydrates used as cold storage and transport materials is provided. Thereafter, the criteria for use of hydrates in refrigeration and related literature are presented. Finally, hydrate applications in refrigeration and air-conditioning are described, including slurry generators and examples of hydrate-based refrigeration architectures.

7.2. Hydrate definition and properties

7.2.1. *Definition of gas hydrates*

Clathrates, from the Greek "klathron" which means closure, are crystalline structures made up of "host" molecules linked together by hydrogen bonds and organized in cages, which enclose "guest" molecules (Jeffrey et al. 1967; Davidson 1973; Jeffrey 1984; Englezos 1993; Sloan 1998; Sloan and Koh 2007). In the case of gas hydrate clathrates or more commonly gas hydrates, the host molecules are water molecules and the guest molecule is a gas molecule (Figure 7.2).

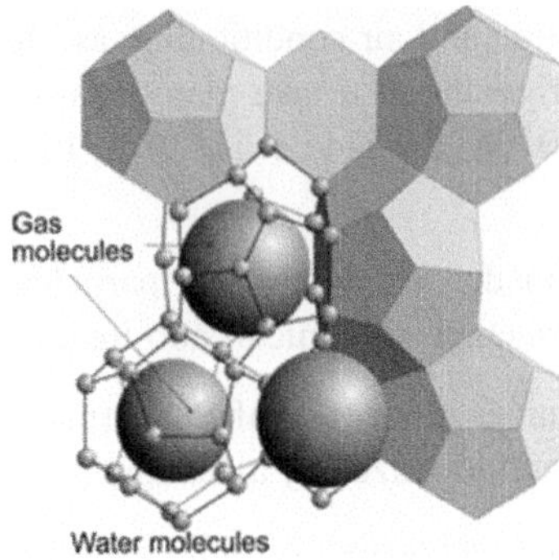

Figure 7.2. *Representation of a gas hydrate (Marum 2009)*

Gas hydrates are crystalline solids that look like ice. The structure of a gas hydrate is stabilized by van der Waals force–type interactions between guest (gas) trapped in the cages and host (water) molecules constituting the cages. When a single type of molecule occupies the different cages, hydrates are called single hydrates and when the cages are occupied by molecules of different natures, they are called mixed hydrates. Gas hydrates are generally formed under conditions of low temperature and pressure above atmospheric pressure (moderate or high).

7.2.2. *Clathrate hydrate structures*

Gas hydrates can be presented as a polyhedral alveolar skeleton. These solids are distinguished by the number and shape of the faces of each polyhedron (cage). For example, cage 5^{12} has 12 pentagonal faces, while cage $5^{12}6^2$ has 12 pentagonal faces and two hexagonal faces. For hydrate structures, there are five different types of cages of varying sizes: 5^{12}, $5^{12}6^2$, $5^{12}6^4$, $4^35^66^3$ and $5^{12}6^8$. These cages are called small (5^{12}), medium ($4^35^66^3$) or large ($5^{12}6^2$, $5^{12}6^4$, $5^{12}6^8$). The cages then organize themselves into three main types of asymmetric three-dimensional structures characterized by the number and size of the cages that compose them. They are cubic I, cubic II and hexagonal H (Sloan 1998; Sloan and Koh 2007), as shown in Table 7.1.

Cubic type I hydrates

Cubic type I hydrates are formed in the presence of small guest molecules such as methane, ethane, carbon dioxide, with diameters between 4.2 and 6 Å. The system is cubic made up of 46 water molecules organized into eight polyhedral cages (two small 5^{12} and six large $5^{12}6^2$) with a unit cell parameter of approximately 12 Å.

Cubic type II hydrates

In type II hydrates, guest molecules with diameters less than 4.2 Å (argon, dioxygen, nitrogen, etc.) occupy the small and large cages. For molecules with a diameter between 6 and 7 Å, only the large cages are occupied (propane, isobutane, tetrahydrofuran (THF)). The system is cubic, with a lattice parameter of approximately 17.3 Å. It can contain mixtures comprising hydrocarbons (HCs) up to pentane.

Hexagonal H type hydrates

Hexagonal type hydrates contain molecules with a diameter between 7 and 9 Å (isopentane or hexane), which occupy the large cages and smaller molecules (methane, hydrogen sulfide, etc.).

Structures	Cubic I		Cubic II		Hexagonal H		
References	(Pauling et al. 1952)		(Claussen 1951)		(Ripmeester et al. 1987)		
Aqueous network diagram Unit cell parameter (Å)*	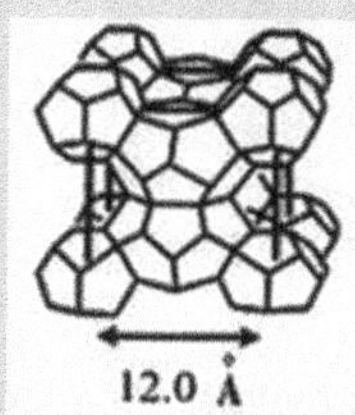 		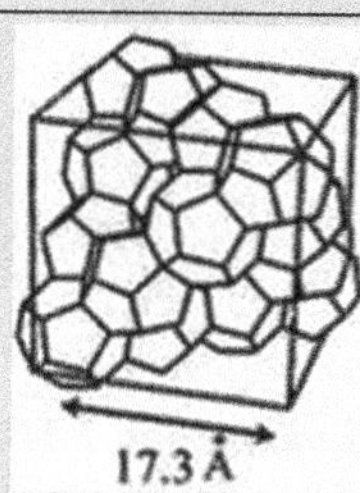 		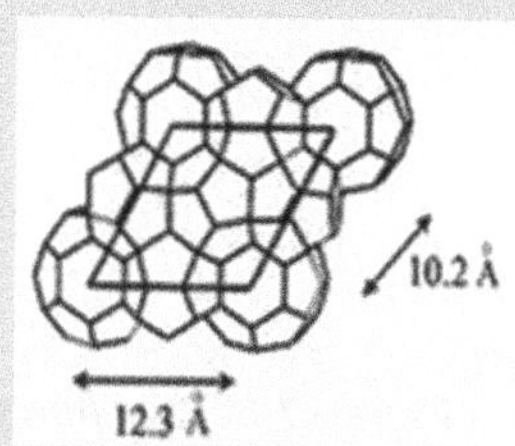 		
Cage j n_j^{mj}**	Small 5^{12}	Large $5^{12}6^2$	Small 5^{12}	Large $5^{12}6^4$	Small 5^{12}	Medium $4^35^66^3$	Large $5^{12}6^8$
Cage Diagram	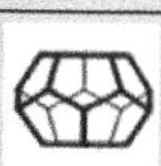	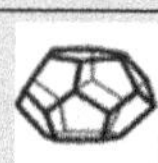	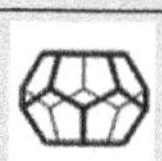		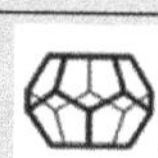	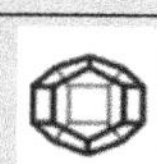	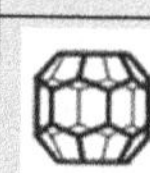
$C_{unit\ cell}$ – H_2O/unit cell	46	136	34				

*Parameter that varies slightly with temperature, studied on structure I (Tse et al. 1987).

**n_j: Number of oxygen atom per face; m_j: number of faces per cage j.

5^{12}: regular dodecahedra.

$5^{12}6^2$: Polyhedra with 14 faces, including 12 pentagons and two hexagons.

$5^{12}6^4$: Polyhedra with 16 faces, including 12 pentagons and four hexagons.

$4^35^66^3$: Polyhedra with 12 faces, including three squares, six pentagons and three hexagons.

$5^{12}6^8$: Polyhedra with 20 faces, including 12 pentagons and eight hexagons.

Table 7.1. *Characteristics of hydrate structures*

7.2.3. *Semiclathrate structures*

The first studies identifying a family of semiclathrate hydrates formed from alkyl-onium salts ($R_4A^+X^-$ with R = methyl, ethyl, propyl, n-butyl, iso-amyl, etc., A = N or P and X = F, Cl, Br, OH, NO_3, etc.) date from the 1940s (Fowler et al. 1940). Later, other authors described their phase equilibria and their structures (McMullan and Jeffrey 1959; Jeffrey and McMullan 1967; Dyadin and Udachin 1984; Nakayama 1987; Lipkowski et al. 2002; Oyama et al. 2005; Shimada et al. 2005; Rodionova et al. 2008). These hydrates are distinguished from the clathrates (sI, sII and sH) because the guest molecules are not only trapped in the cages, but participate in the structure itself. In fact, the guest molecule is a salt whose anions are involved in the water lattice structure by hydrogen bonding, the host cavities being only occupied by the cation. They are thus called "semiclathrates".

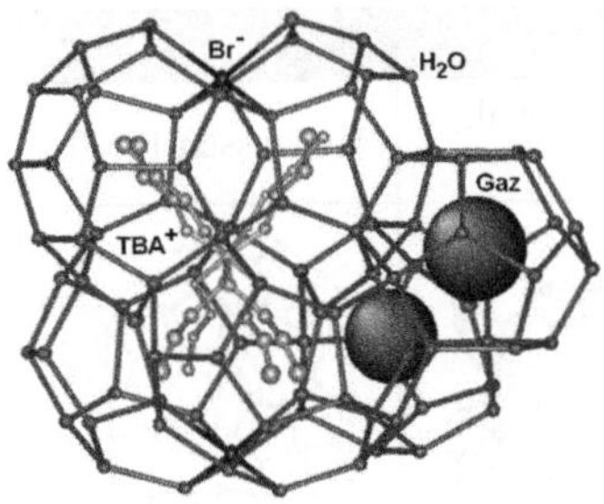

Figure 7.3. *Structure of a mixed gas + TBAB hydrate (Shimada et al. 2005). For a color version of this figure, see www.iste.co.uk/odru/heat1.zip*

In addition, salt (TBAB) semiclathrate hydrates can also trap gas molecules (H_2, CH_4, CO_2) in the small, usually empty, cages (5^{12}) of their structure (Shimada et al. 2005), leading to a mixed gas + salt hydrates (Figure 7.3).

7.2.4. *Hydration number*

The composition of a hydrate is generally represented by the formula G.nbhH2O (G: guest molecule, i.e., gas and salt). The chemical equation for the formation/dissociation of the hydrate is then written as:

$$G + nb_h\, H_2O \rightleftarrows G.nb_hH_2O \text{ (hydrate)} \quad [7.1]$$

The structure of a hydrate determines its composition, represented by its hydration number nb_h, which is the average number of moles of water per mole of guest molecule.

$$nb_h = \frac{n_{\mathrm{H_2O\ in\ hydrate}}}{n_{\mathrm{G\ in\ hydrate}}} \qquad [7.2]$$

The ratio between the number of water molecules and the number of cavities gives a lower limit to nb_h, corresponding to the ideal (stoichiometric) case where all the cavities would be occupied by the guest molecules. In fact, "gas" hydrates are non-stoichiometric compounds: they are stable without all the cages being occupied. The occupancy rate of the cavities then depends on pressure and temperature (Sloan 1998) and corresponds to a real hydration number higher than the ideal hydration number. For example, the hydration number of structure I CO_2 hydrate varies in practice between 6.07 and 8.69 (Larson 1955; Parrish and Prausnitz 1972; Vlahakis et al. 1972), while the ideal hydration number of structure I is equal to 5.75 (= 46/8, i.e., 46 water molecules involved in the structure of eight cavities). Moreover, hydrates formed from liquid or salt, such as THF hydrates or TBAB semiclathrate hydrates, are stoichiometric compounds.

7.2.5. *Phase diagram of water-CO_2 mixtures including CO_2 hydrates*

From theoretical or experimental data (often obtained by calorimetry), it is possible to represent the thermodynamic conditions of existence of hydrates in a phase diagram. For gas hydrates, (P, T) representation is usual. In the case of CO_2 hydrates, the three-phase liquid water–hydrate–vapor equilibrium domain (L_w-H-V) extends in temperature from 271.8 to 283.1 K and in pressure from 1.0 to 4.5 MPa. Figure 7.4 represents the phase diagram for a water–CO_2 mixture at temperatures between 263 and 293 K and pressures between 0 and 6 MPa. This diagram was established from data collected in the literature (Deaton and Frost 1946; Unruh and Katz 1949; Larson 1955; Robinson and Mehta 1971; Vlahakis et al. 1972; Adisasmito et al. 1991; Yang et al. 2000). Several domains, curves and points can be distinguished:

– five two-phase equilibrium domains: liquid water–vapor (L_w-V); liquid water–liquid CO_2 (L_w-L_{CO2}); ice–vapor (I-V); ice–hydrate (I-H); and liquid water–hydrate (L_w-H);

– six three-phase equilibrium curves: liquid water–liquid CO_2–vapor (L_w-L_{CO2}-V); liquid water–liquid CO_2–hydrate (L_w-L_{CO2}-H); liquid water–hydrate–

vapor (L_w-H-V); ice–hydrate–vapor (I-H-V); ice–liquid water–vapor (I-L_w-V); and ice–liquid water–hydrate (I-L_w-H);

– two quadruple points: Q_1 (271.8 K, 1.0 MPa) and Q_2 (283.1 K, 4.5 MPa) representing the coexistence of four phases: ice–hydrate–liquid water–vapor (I-H-L_w-V) for Q_1; hydrate–liquid water–vapor–liquid CO_2 (H-L_w-V-L_{CO2}) for Q_2.

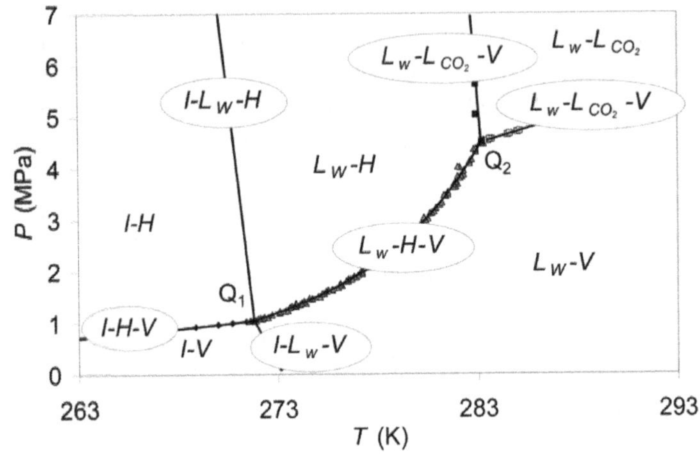

Figure 7.4. *Phase diagram (P, T) of water–CO2 mixture*

For hydrate slurry use, L_w-H-V phase equilibrium is interesting where CO_2 is in vapor form, dissolved and trapped in the hydrate, while the water is in the liquid state or forms the hydrate structure.

7.3. Hydrate systems for cold storage and distribution

Discovered by Sir Humphry Davy in 1811, clathrate hydrates have long been studied academically (Davy 1811b, 1811a; Faraday 1823). Research on natural gas hydrates intensified from the 1930s when the oil and gas industry sought to inhibit their formation in pipelines (Hammerschmidt 1934; Deaton and Frost 1946). The research on hydrates applied to refrigeration started in the 1980s. Several compounds were considered to form hydrates, including refrigerant gases, HCs, CO_2 or quaternary salts. These hydrates, with single or mixed compositions, mostly clathrate or semi-clathrate structures, studied for cold storage and/or distribution, are presented in this section (Delahaye et al. 2018).

7.3.1. *Refrigerant gas hydrate applied to cold storage*

The first studies on hydrates for refrigeration purposes aimed at developing thermal energy storage systems based on refrigerant gas hydrates, having clathrate structure and thermodynamic properties suitable for refrigeration applications (high dissociation enthalpy, low pressure and wide temperature range). Thus, various authors studied gas hydrates in heat pumps or direct-contact crystallizers using water and refrigerant gases, for example, CFC-11 (Akiya et al. 1987, 1988), CFC-12 (Ternes 1984; Carbajo 1985; Mori and Mori 1989a) and HFC-134a (Mori and Mori 1989b). The objective was to inject a refrigerant fluid into water stored in a crystallizer (a heat exchanger such as an evaporator or a vessel) to exchange directly heat with water and form clathrates during the cooling operation; the clathrate could be melted later to produce chilled water in an air-conditioning system.

Research on various refrigerant gas hydrates used as cold storage material continued until the late 1990s (Mori and Isobe 1991; Isobe and Mori 1992; Utaka et al. 1993; Tanii et al. 1996a, 1996b; Akiya et al. 1997a, 1999) and then focused on two main refrigerant fluids until today, that is, HCFC-141b (Tanii et al. 1996b, 1997; Fu et al. 2001; Liang et al. 2002; Liu et al. 2003; Bi et al. 2004, 2006, 2009, 2010, 2012, 2014; Li et al. 2004, 2005; Xie et al. 2005a, 2005b, 2005c, 2010; Wang et al. 2008) and HFC-134a (Li et al. 2006; Wu et al. 2012; Ahmed and Gadalla 2013; Gadalla and Ahmed 2013; Zafara et al. 2015). Among these studies, there may be mentioned those conducted on HCFC-141b gas hydrate in glass tubes (Li et al. 2004) or in crystallizers (Bi et al. 2004) for cold-storage purposes, while Wang et al. (2008) used HCFC-141b as a substitute to study the flow characteristics and the formation by cooling a hydrate slurry in a circulating loop. In addition, a novel refrigeration system was proposed based on the use of refrigerant gas hydrate slurries, formed by mixing HCFC-22 and water or HFC-32 + cyclopentane and water (Ogawa et al. 2006). The use of this mixture would indeed limit the environmental impact of refrigerants.

7.3.2. *CO_2 hydrates applied to cold storage and distribution*

The idea of using environmentally friendly gas to form clathrate hydrates applied to air-conditioning was first proposed in the late 1990s (Matsuo et al. 1999). The objective was to replace CFC-type and HFC-type refrigerants by refrigerants with "zero" ozone-depleting potential (ODP) and minimal global warming potential (GWP) and it was concluded that propane, isobutene and CO_2 hydrates were thermodynamically viable for a cold storage system. Fournaison et al. (2004) also proposed studying CO_2 hydrate as PCM for refrigeration applications because of its high dissociation enthalpy (374 $kJ \cdot kg^{-1}$: higher than that of ice of 333 $kJ \cdot kg_{water}^{-1}$). The use of CO_2 hydrate slurries was then considered by various authors for both cold storage or distribution (Marinhas et al. 2006, 2007; Delahaye et al. 2008, 2011; Jerbi et al. 2010a, 2013; Choi et al. 2014, 2015; Wang et al. 2014, 2015b; Xie et al. 2014; Zhou et al. 2015).

CO_2 hydrate slurry production has the advantage of being achievable based on CO_2 injection in a pre-cooled aqueous solution (Marinhas et al. 2006, 2007), on the same principle as direct-contact crystallizers using water and refrigerant gases. Such non-mechanical process avoids the use of scraped or brushed surface heat exchangers, usually power limited by mechanical parts, as in the case of ice slurries. Similarly to refrigerant gas clathrate hydrates (Watanabe et al. 2005; Okutani et al. 2007; Li et al. 2008), kinetic approaches based on the use of additives such as SDS or nanoparticles were also carried out to optimize and control CO_2 hydrate formation/dissociation rate (Delahaye et al. 2011; Egenolf-Jonkmanns et al. 2012; Choi et al. 2014; Fischer et al. 2014; Lin et al. 2014; Youssef et al. 2014; Sun and Kang 2015). In addition, Yang et al. (2008) used mixtures of CO_2 and argon, or helium, to continuously produce CO_2 clathrate hydrate slurries and found that a vigorous mixing was favorable to hydrate formation kinetics. This study, originally planned for CO_2 capture process, is also useful for CO_2-hydrate-based cold storage and transport applications.

7.3.3. *Quaternary salt hydrates for cold storage and distribution*

In the late 1990s, other hydrate structures, namely semi-clathrate hydrates of quaternary salts, in particular TBAB hydrates, were also considered for the first time for cold storage and transportation applications (Fukushima et al. 1999). These authors produced many studies and patents on TBAB hydrate slurry properties (Hayashi et al. 2000; Takao et al. 2001, 2002, 2005, 2007; Tanasawa et al. 2002; Ogoshi and Takao 2004). Unlike gas hydrates, semi-clathrate hydrates of quaternary salts have the advantage of being stable at atmospheric pressure (Fowler et al. 1940; Jeffrey et al. 1967; Dyadin and Udachin 1984; Nakayama 1987). According to Oyama et al. (2005), TBAB hydrate has at least two crystalline structures, commonly named type A and B, corresponding to hydration numbers of 26 and 38, respectively. Other hydration numbers for various quaternary salt hydrates are provided in a review (Zhang et al. 2010b).

In fact, many investigations were carried out on semiclathrate hydrates applied to refrigeration, but most were dedicated to TBAB hydrates, in particular their structural, thermophysical and rheological properties (Lipkowski et al. 2002; Darbouret et al. 2005; Oyama et al. 2005; Xiao et al. 2006; Song et al. 2009; Li et al. 2010; Ma et al. 2010; Zhang et al. 2010a, 2013, 2014; Kumano et al. 2011a, 2011b;Ma and Zhang 2013 Shi et al. 2014b, 2016; Youssef et al. 2014; Zhou et al. 2014; Rondeau-Mouro et al. 2015; Wang et al. 2015a, 2016; Mahmoudi et al. 2016), but also their implementation in cooling systems (Daitoku and Utaka 2007, 2010; Douzet et al. 2013; Shi et al. 2013, 2014a, 2014b). TBAB hydrate slurries can be obtained by cooling an aqueous solution of TBAB. Takao et al. (2007) suggested to use fine nucleating particles in the solution to reduce the supercooling degree. As ice crystals, hydrate crystals can nucleate and grow on the heat exchanger walls, and then form a hydrate layer which deteriorates the heat transfer performance.

Darbouret et al. (2005) used brushed surface heat exchangers to produce TBAB hydrate slurries, or studied different scraper shapes and the force required to remove TBAB hydrate particles from the heat transfer surface and contribute to their fluidization (Daitoku and Utaka 2007, 2010). Different methods of TBAB hydrate slurry generation, as continuous and discontinuous cooling, or supercooling approaches, were also studied to limit hydrate crystal adhesion to the heat exchanger wall (Shi et al. 2013, 2014a, 2014b).

Other salts, such as TBAF, TBAC, TBAN, TBAAc, TBPB, TBPC and TBMAC (Mayoufi et al. 2010, 2011, 2012; Hashimoto et al. 2011; Sakamoto et al. 2011; Clain et al. 2012; Suginaka et al. 2012; Lin et al. 2013; Clain et al. 2015; Wang et al. 2015a; Babu et al. 2016), were also investigated because of their ability to form semi-clathrate hydrates suitable for cooling applications. Finally, recent studies were conducted on mixtures of salts involving TBAB and TBAC (Sato et al. 2013; Shi et al. 2016a), TBAB and TBPB (Lin et al. 2014), and TBAF, TBAB, and TBPB (Wang et al. 2016).

7.3.4. *Other hydrates applied to cold storage and distribution*

HC clathrate hydrates are generally studied for oil and gas industry concerns, but they were also investigated for refrigeration purposes, as the early work on propane hydrates (Matsuo et al. 1999). Hydrate-forming cyclopentane-in-water emulsions (Nakajima et al. 2008; Chami et al. 2021) or isobutane hydrates (Xie et al. 2009) were also examined for thermal energy storage media in air-conditioning systems.

Another compound, THF, often used as a thermodynamic additive, was also considered from the 1990s to form single THF hydrate to be used as a cold storage medium (Akiya et al. 1997b). Kinetic studies showed that THF hydrate formed outside of a heat transfer tube has a similar subcooling degree and induction time as those of ice (Xie et al. 2008), while accelerated nucleation of THF hydrate was possible in the presence of zeolite (Wang et al. 2012).

Finally, trimethylolethane (TME) hydrate slurries were also studied as thermal storage material (Yamazaki et al. 2002; Kakiuchi et al. 2003) and latent heat transport medium (Suzuki et al. 2006, 2009). According to the work of Yamazaki et al. (2002), TME forms clathrate-type hydrate with three water molecules, initially called trihydrate. TME hydrate structure is thus different from classical clathrate hydrate structures (I, II and H), but presents a melting point at 302.8 K and a heat of melting of 218 kJ/kg, comparable to those of semi-clathrate hydrates of quaternary salts. Moreover, TME hydrates are stable at atmospheric pressure and can be obtained by cooling an aqueous solution of TME in a tank (Suzuki et al. 2006). A review on PCMs (Zhang et al. 2010b) presents the thermophysical properties of TBAB, TBAC, TBAF and TME hydrate slurries.

7.3.5. *Mixed gas-salt/THF hydrates for cold storage and distribution*

In order to reduce equilibrium pressures of CO_2 hydrates, several studies were carried on the CO_2-water mixture in the presence of various thermodynamic additives. Based on work applying hydrates to the separation of CO_2/N_2 mixtures (Kang et al. 2001), THF was used to form CO_2-containing hydrates at moderate pressure for cold storage purposes (Delahaye et al. 2006; Marinhas et al. 2007; Martínez et al. 2008; Mayoufi et al. 2010, 2011; Choi et al. 2014; Sun and Kang 2015).

In addition, the discovery of the ability of salt hydrates to trap gas molecules (Shimada et al. 2005) triggered numerous studies on semiclathrates. The first were devoted to mixed hydrates of gas and TBAB (Hashimoto et al., 2006; Arjmandi et al., 2007; Chapoy et al., 2007; Li et al., 2007; Hashimoto et al., 2008; Lin et al., 2008) for various applications (gas separation/storage, energy storage).

Other salts such as TBPB were considered to form mixed salt-gas hydrates applied to cold storage (Lin et al. 2008; Mayoufi et al. 2010, 2011; Lin et al. 2014; Youssef et al. 2014). The interest of using THF/salt and CO_2 was to form high-energy CO_2-containing hydrates more stable than single CO_2 hydrates and thus easier to implement in cooling processes. Adding salt/THF thus allowed pressure reduction of approximately 70–90% compared to single CO_2 hydrates.

7.4. Criteria for use of hydrates in refrigeration

An indirect refrigeration system has an environmental interest if the secondary refrigerant fluid used has little or no impact on the environment, as is the case of most hydrate slurries. However, other essential criteria must be fulfilled to consider an application of hydrate slurry for cold storage and transportation:

– Thermodynamic criterion: hydrates must have suitable equilibrium conditions and high dissociation enthalpy.

– Flow criterion: hydrate slurries must have suitable flowing conditions with low-pressure drop to limit pumping power.

– Thermal criterion: hydrate systems must have a good ability to exchange heat with the medium to cool.

– Kinetic criterion: hydrate crystallization (fraction, size) and dissociation rate must be controlled.

– Energy criterion: hydrate-based cooling applications must have a good coefficient of performance (COP).

– Environment criterion.

Thermodynamic, flow, thermal and kinetic criteria and related literature studies on hydrates applied to refrigeration are presented in this section. Moreover, a short review of energy studies related to hydrates in cooling applications is presented below.

7.4.1. *Thermodynamic criterion*

The basic criterion to be considered for hydrate-based cooling application is thermodynamics. First of all, hydrates must have a wide temperature range consistent with the intended applications (cooling process, warehouse storage, air conditioning, etc.) and be able to run at stable temperatures that reduce the thermodynamic irreversibility of the primary cold production system (evaporator) and ensure temperature uniformity for the user. Moreover, hydrates must have a high phase change enthalpy allowing large amounts of energy to be stored, carried, and released. In addition, in the case of gas hydrates, a low equilibrium pressure is recommended for the technical and economic viability of the process.

Calorimetric tools, such as a differential scanning calorimeter (DSC) or differential thermal analysis (DTA) device (Dalmazzone et al. 2017), were often used to obtain these data. Hydrate–liquid–vapor equilibrium points (H-Lw-V) and hydrate dissociation enthalpy can be directly measured by applying hydrate formation and dissociation cycles.

Table 7.2 lists thermodynamic properties of various PCMs suitable for refrigeration purposes, including clathrate and semiclathrate hydrates.

Type/structure	PCM/hydrate	T (K)	ΔH (kJ·kg^{-1})	P (kg·m^{-3})	nbh	Reference
Pure paraffin	Tetradecane	278.9	227	763	–	(Bo et al. 1999)
	Pentadecane	283.0	206	769	–	
	Hexadecane	291.2	236	773	–	
Paraffin + acid	Pentadecane + Acides	286.4	142	–	–	(Zalba et al. 2003)
Paraffin blend	C15-C16	281.1	153	–	–	(Abhat 1983)
	RT6	281.1	174	770	–	(Huang et al. 2010)
	RT10	283.1	152	780	–	
	RT20	295.1	130	750	–	
Ice	Water	273.1	333	917	–	(Weast et al. 1987)
Clathrate SI	CO_2	271.8–283.1	374	1,065	7.23	(Kang et al. 2001; Delahaye et al. 2006; Marinhas et al. 2006)
Clathrate SII	HCFC-141b	268.4–281.5	330	–	17	(Liang et al. 2001, 2002)
	Propane	278.8	346	–	17	(Matsuo et al. 1999)
	Iso-butane	275.03	339	–	17	
	THF	277.5	260	999	17	(Dyadin et al. 1973; Hanley et al. 1989; Otake et al. 2000; Delahaye et al. 2006)

Semiclathrate	TBAB	285.1 9.5–12.4	193.2–205	1,030–1,082	24–38	(Lipkowski et al. 2002; Ogoshi and Takao 2004; Darbouret et al. 2005; Oyama et al. 2005)
	TBAF	27.2–28.3	229.7	1,035–1,057	28.6–32.3	(Dyadin and Udachin 1984; Nakajima et al. 2008; Mayoufi et al. 2010)
	TBAC	14.7–15	200.7	1,029–1,034	29.7–32.1	
	$TBANO_3$	278.4	175	–	26	(Dyadin and Udachin 1984; Mayoufi et al. 2010)
	TBPB	281.1–282.4	203.8	1,140	32–36.6	(Dyadin and Udachin 1984; Mayoufi et al. 2010; Clain et al. 2012; Suginaka et al. 2012; Lin et al. 2013)
Trihydrate	TME	30	218	1,090–1,120	3	(Yamazaki et al. 2002)

Table 7.2. *Thermodynamic properties of PCMs suitable for secondary refrigeration; T: melting temperature; ΔH: melting enthalpy; ρ: density; nbh: hydration number*

According to Table 7.2, several hydrates exhibit dissociation enthalpy higher than that of ice. CO_2 hydrate has the highest dissociation enthalpy and therefore can be considered as having high potential PCM. However, refrigerant gas (HCFC-141b) hydrate has also a high heat content, which explains its popularity since the 1980s, especially as refrigerant hydrates are often stable at low pressure. However, currently, the environmental constraints on refrigerant gases limit the applicability of these hydrates. Likewise, propane and iso-butane hydrates have good energy properties but their flammability is an issue. THF hydrate has also a high dissociation enthalpy, although less than that of ice, but due to its hazardous nature, it is not suitable for refrigeration and air-conditioning applications.

Salt hydrates have meanwhile lower enthalpies than those of gas and THF clathrates, about 200 kJ·kg^{-1}, but they are more interesting than conventional paraffin mixtures. The advantage of salt hydrates lies in their stability at atmospheric pressure, which facilitates their implementation for storage and cold transportation applications.

The limitation of CO_2 hydrate formation pressures has also motivated the studies of their thermodynamic properties. Based on work related to mixed hydrates applied to gas separation (Kang et al. 2001), different CO_2-containing systems were characterized for refrigeration purposes. These systems have much lower equilibrium pressures than simple CO_2 hydrates, while maintaining a high dissociation enthalpy. Two types of systems have been studied: the first is a THF-based system model, and the second is based on alkyl-onium salts. Some data related to these two systems are shown in Figures 7.5 and 7.6.

Knowledge of hydration number (ratio between the number of moles of water and guest molecules in hydrate) and therefore hydrate composition is also essential to control hydrate fraction in cold storage and distribution processes. These data can be obtained by experimental methods such as X-ray diffraction (Udachin et al. 2001), Raman spectroscopy (Uchida et al. 1995), nuclear magnetic resonance (Ripmeester and Ratcliffe 1998) or PVT measurements in reactor (Kang et al. 2001).

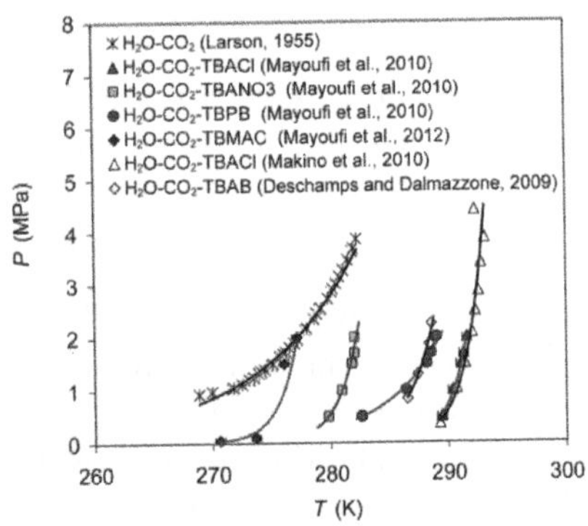

Figure 7.5. *Lw-H-V equilibrium conditions of water-CO_2, water-CO_2-salt (with salt: TBMAC, TBACl, $TBANO_3$, TBPB or TBAB). For a color version of this figure, see www.iste.co.uk/odru/heat1.zip*

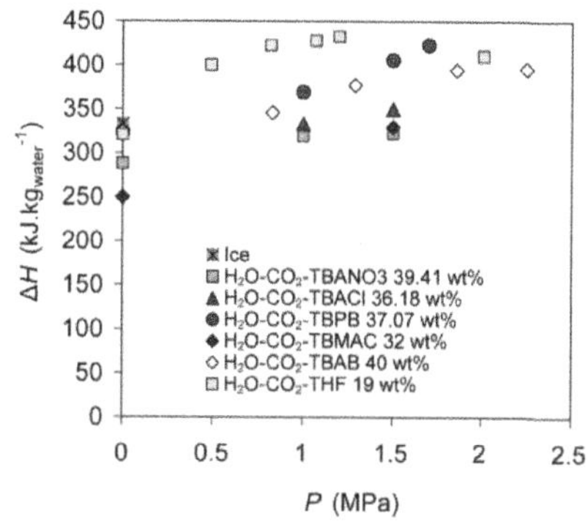

Figure 7.6. *Dissociation enthalpies of mixed hydrates of CO_2 + additive (with additive: TBMAC, $TBANO_3$, TBPB, TBACl, TBAB, THF) in $kJ{\cdot}kg_{water}^{-1}$. For a color version of this figure, see www.iste.co.uk/odru/heat1.zip*

From equilibrium data and dissociation enthalpies, it is also possible to determine hydration number (Mayoufi et al. 2010). The principle is to express hydrate dissociation enthalpy in two different units: one expressed in kJ per mole of gas in the hydrate, deduced from the equilibrium curves (T, P) hydrate–liquid–vapor (Lw-H-V), the other in kJ per mole of water in the hydrate, directly measured by calorimetry. The hydration number (from equation [7.2]) is equal to the ratio of enthalpies:

$$nb_h = \frac{n_{\text{water in hydrate}}}{n_{\text{gas in hydrate}}} = \frac{\Delta H\left(\text{kJ. mol}_{\text{gas}}^{\ -1}\right)}{\Delta H\left(\text{kJ. mol}_{\text{water}}^{\ -1}\right)} \qquad [7.3]$$

7.4.2. *Flow criterion*

Besides a high dissociation enthalpy and stability conditions appropriate to refrigeration application, the capacity of hydrate slurries to convey energy and transfer it to users is a key factor. The objective is to circulate sufficiently concentrated hydrate slurries (solid fraction of 20–30%) with a suitable flow rate, both to limit pressure drops and pumping power and to prevent the risk of sedimentation or plugging.

In order to implement such systems, hydrate slurry flow conditions must be controlled. In particular, it is necessary to understand the influence of the hydrate particles dispersed in the liquid phase on flow properties. Depending on flow conditions, the pressure drops induced by fluid flow are controlled either by rheological characteristics such as apparent viscosity in laminar regime or by friction factor in turbulent regime (Sinquin et al. 2004).

Rheology is used to determine the shear rate $\dot{\gamma}$ (s^{-1}) of materials (as hydrate slurries) subjected to a shear stress τ (Pa), and thus allows fluid flow properties to be determined. Fluid flow behavior is generally classified into Newtonian or non-Newtonian fluids. For Newtonian fluids such as water, the shear rate is proportional to the shear stress: the proportionality factor (constant) is called the dynamic viscosity μ (Pa·s). In the case of non-Newtonian fluids, the shear rate and the shear stress are not proportional. The behavior of non-Newtonian fluid can be described by using the general Herschel–Bulkley equation:

$$\tau = k\dot{\gamma}^n + \tau_0 \quad [7.4]$$

where k is the fluid consistency coefficient ($Pa \cdot s^n$), τ_0 is the minimal shear stress (also called yield stress) and n is the behavior index. If $n = 1$ and $\tau_0 = 0$, then the fluid is Newtonian. The more n differs from 1, the more the fluid is non-Newtonian. More generally, fluid behaviors can be mainly classified into various types: pseudo-plastic (shear thinning), dilatant fluid (shear thickening), Bingham and Herschel–Bulkley fluid. There are also the Ellis, Casson, and Sisko models or the Shangraw equation.

The relationship between the shear rate $\dot{\gamma}$ and the shear stress τ is generally obtained by using one of these two experimental methods: a rotary viscometer or a capillary viscometer (also called Ostwald viscometer).

One of the first rheological tests on hydrate slurries (Pinder 1964) was performed using a rotary viscometer on a suspension of hydrogen sulfide hydrates and THF (H_2S-2THF-17H_2O).

Later, various rheological studies were carried out to address the problems of agglomeration and transportability in pipelines. These studies were devoted to natural gas hydrates in suspension in a carrier phase consisting of HCs (Austvik and Bjorn 1992; Austvik et al. 2000; Andersson and Gudmundsson 1999; Camargo et al. 2000; Camargo and Palermo 2002; Fidel-Dufour and Herri 2002; Fidel-Dufour et al. 2006; Peysson et al. 2003, 2004; Sinquin et al. 2004; Cameirao et al. 2011; Gainville et al. 2011; Peng et al. 2012; Webb et al. 2012, 2013, 2014; Yan et al. 2014; Zylyftari et al. 2014; Pandey et al. 2017) or in an aqueous carrier phase (Nygaard 1989; Andersson and Gudmundsson 2000; Pandey et al. 2017).

Hydrate slurry flow properties were also studied for refrigeration applications, with rheological work on aqueous suspensions of salt hydrates (Fukushima et al. 1999; Darbouret et al. 2005; Xiao et al. 2006; Ma et al. 2010; Hashimoto et al. 2011; Kumano et al. 2011a; Clain et al. 2012) or gas hydrates (Delahaye et al. 2008, 2011; Wang et al. 2008; Jerbi et al. 2013; Oignet et al. 2017a). Table 7.3 includes rheological studies on hydrate slurries in various fields including refrigeration, oil and gas industry, and gas separation, as well as fundamental work (Pinder 1964; Oyama et al. 2002).

Hydrate	Liquid	Viscometer	ϕ_s	Behavior	References
H_2S + THF	Aqueous	Rotational	<0.01	Thixotropic gel μ_{app}: from 60 to 23 mPa·s	(Pinder 1964)
HC	Organic	Rotational	–	No measure of τ_w	(Austvik and Bjorn 1992)
TBAB	**Aqueous**	**Capillary**	**0.22-0.31**	**OdW S-Thin.: μ_{app}: 30–2,000 mPa·s**	**(Fukushima et al. 1999)**
HC	Organic + AA	Capillary	0–0.1	Bingham: k : 3.4–5.5 mPa·s	(Andersson and Gudmundsson 1999)
CH_4	Aqueous	Capillary	0.01–0.1	Bingham: k: 1–3.5 mPa·s	(Andersson and Gudmundsson 2000)
HC	Organic	Rotational	0–0.5	HB S-Thin.: μ_{app}: 10-1200 mPa.s	(Camargo et al. 2000)
CO_2	Aqueous	Magnetic (stress)	–	μ_{app} increase before nucl. then decrease	(Oyama et al. 2002)
HC	Organic + TA	Capillary	–	μ_{app} increase after nucl. then decrease	(Fidel-Dufour and Herri 2002)
HC	Organic	Capillary	0.1–0.3	OdW S-Thicken. $n = 2$, $k = 2 \times 10^{-3}$ mPa·s	(Peysson et al. 2003)
TBAB	**Aqueous**	**Capillary**	**0.04–0.53**	**Bingham: k: 8–170 mPa·s**	**(Darbouret et al. 2005)**
TBAB	**Aqueous**	**Capillary**	**0–0.16**	**OdW S-Thin.: μ_{app}: 4–42 mPa·s**	**(Xiao et al. 2006)**
CH_4	Organic + AA	Capillary	0.07–0.18	Newtonian: 2.5–3.5 mPa·s	(Fidel-Dufour et al. 2006)
CO_2	**Aqueous**	**Capillary**	**0.04–0.1 0.1–0.2**	**OdW S-Thicken. HB S-Thin.: μ_{app}: 4–42 mPa·s**	**(Delahaye et al. 2008)**

R141b	**Aqueous**	**Capillary**	**0.1–0.68**	**OdW S-Thicken.: μ_{app} : 1.1–1.7 mPa·s (at 400 s^{-1})**	**(Wang et al. 2008)**
TBAB	**Aqueous**	**Capillary**	**0.06–0.2**	**OdW S-Thin.: μ_{app}: 3–100 mPa·s**	**(Ma et al. 2010)**
CO_2	**Aqueous + TA**	**Capillary**	**0.04–0.1**	**Newtonian: μ_{app}: 3.3–16.6 mPa·s**	**(Delahaye et al. 2011)**
TBAB	**Aqueous**	**Ubbelohde**	**0.02–0.25**	**OdW S-Thin.: μ_{app}: 2–5 mPa·s**	**(Kumano et al. 2011a)**
TBAB	**Aqueous**	**Plate**	**0.12–0.7**	**OdW S-Thin.: μ_{app}: 3.5–1,000 mPa·s**	**(Hashimoto et al. 2011)**
TBAF	**Aqueous**	**Plate**	**0–0.42**	**OdW S-Thin.: μ_{app}: 10–750 mPa·s**	
HC	Organic	Capillary	0.1–0.5	μ_{app}: 5.2–9.6 mPa·s	(Cameirao et al. 2011)
HC	Organic	Capillary	0.05–0.30	μ_{app}: 32–220 mPa·s	(Gainville et al. 2011)
CH_4/CO_2	Aqueous	Capillary	0–0.12	Non-Newtonian: μ_{app}: 1.26–4.8 mPa·s	(Waycuilis et al. 2011)
TBPB	**Aqueous**	**Capillary**	**0–0.28**	**OdW S-Thin.: μ_{app}: 4–41 mPa·s**	**(Clain et al. 2012)**
HC + Esters polymers	Organic	Capillary	0.05–0.30	S-Thin.: μ_{app}: 4–70 mPa·s	(Peng et al. 2012)
HC	Organic	Rotational	0.2–0.45	HB S-Thin.: μ_{app}: 500–3,000 mPa·s	(Webb et al. 2012)
CO_2	**Aqueous**	**Capillary**	**0–0.21**	**OdW S-Thin.: μ_{app}: 3–10 mPa·s**	**(Jerbi et al. 2013)**
HC	Organic	Rotational	0.1–0.3	HB S-Thin.: μ_{app}: 1,000–10,000 mPa·s	(Webb et al. 2013)
HC	Organic	Rotational	0.1–0.4	S-Thin.: μ_{app}: 100–2,000 mPa·s	(Webb et al. 2014)
HC	Organic	Capillary	0.06–0.35	HB S-Thin.: μ_{app}: 7–16 mPa·s	(Yan et al. 2014)
Cyclopentane	Organic	Rotational couette	0–0.15	OdW S-Thin μ_{app}: 0.1–1,000 mPa·s	(Zylyftari et al. 2014)
HC	Aqueous organic	Rotational couette	–	μ_{app}: 23–27 mPa·s (at 100 s^{-1})	(Pandey et al. 2017)
CO_2	**Aqueous + SDS**	**Capillary**	**0–0.155**	OdW S-Thin μ_{app}: 7–21 mPa·s	**(Oignet et al. 2017a)**

Table 7.3. *Work on the rheology of hydrate slurry; Bold: work on refrigeration applications; HC: hydrocarbon; TA: surfactant; AA: anti-agglomerant; ϕ_s: hydrate volume fraction; R141b: HCFC 141-b refrigerant; ODW: Ostwald-de Waele; S-Thin.: shear thinning; S-Thicken.: shear thickening; HB: Herschel–Bulkley*

According to Table 7.3, several key features of hydrate slurry rheology can be observed in literature:

– Various viscometers are used, such as rotating, magnetic or plate, but the most common method applied to hydrate slurries is the capillary viscometer, well adapted to dynamic loop geometry.

– HC hydrates are studied in organic phase, while salt, CO_2 and also CH_4 hydrate studies are carried out in aqueous phase.

– Hydrate slurry viscosity increases with solid particle fraction, as do most slurries (ice, paraffin, etc.).

– Most hydrate slurries are non-Newtonian even at low solid fraction, unlike the ice slurry (Ayel et al. 2003).

– Differences on rheological models (Bingham, Ostwald-de Waele, Herschel-Bulkley, shear thinning, shear thickening, thixotropic, etc.) and viscosity values are noticed. This could be due to differences in composition of the carrier phases (organic vs. aqueous, with vs. without surfactant – such as SDS – or anti-agglomerant additives, salt vs. gas dissolved) and hydrates (salt vs. gas), as well as differences in hydrate–liquid interactions; differences in methodological approaches, although applied on the same type of viscometers, may also explain discrepancies.

– Despite differences in rheological models (with or without yield stress, behavior index lower than or equal to 1), hydrate slurries are usually shear-thinning/pseudoplastic (viscosity decreasing with increasing shear rate) Ostwald-de Waele or Herschel-Bulkley, and sometimes Bingham, which is close to shear-thinning behavior over a limited range of shear rate, as reported on ice slurries (Ayel et al. 2003).

Finally, as highlighted in the literature review on ice slurry (Ayel et al. 2003), the yield stress cannot be regarded as a fundamental property of the material since its value depends on the model or the viscometer. The disparity of rheological models could be linked to the empirical nature of the yield stress.

7.4.3. *Thermal criterion*

The interest of hydrate slurry is to store and transport large amounts of thermal energy, but it is also necessary to control heat transfer during the stages of hydrate formation and dissociation, in particular to transfer the cold to users through heat exchangers. Heat transfer characteristics of TBAB hydrate slurries were thus studied in literature by various authors (Song et al. 2009; Ma et al. 2010; Kumano et al.

2011b; Ma and Zhang 2013; Shi et al. 2014b; Zhang et al. 2014; Zhou et al. 2014; Shi and Zhang 2016; Wang et al. 2016).

It was, for example, pointed out that the local heat transfer coefficient of type B TBAB hydrate slurry was larger than that of type A under the similar thermal conditions, implying that type B TBAB hydrate slurry was more favorable to be used as secondary refrigerant (Zhang et al. 2014). Heat transfer characteristics were also determined for CO_2 hydrate slurries (Yang et al. 2008; Oignet et al. 2017b). It was found that a heat transfer coefficient from a He/CO_2 gas mixture was at least 50% higher than that obtained from a comparable Ar/CO_2 mixture (Yang et al. 2008). Moreover, except for micro-encapsulated PCM, CO_2 hydrate slurries provide higher heat transfer coefficient than that of other slurries (Oignet et al. 2017b), with comparable solid fraction and flow regimes (Table 7.4). Heat transfer was also studied in the cold storage process of refrigerant HCFC-141b gas hydrates, in the presence of additives (such as SDS) to enhance heat storage and release processes (Xie et al. 2005b, 2010; Bi et al. 2009). Table 7.4 includes thermal studies on the convective heat transfer coefficient (CHTC) for various slurries in refrigeration.

System	Reynolds number	Solid fraction	CHTC	Experimental apparatus	Reference
Micro-encapsulated PCM	Re = 2,320–3,503	5 wt.%	5,000–6,500	Horizontal circular pipe	(Wang et al. 2007)
Stabilized ice slurry	Re = 2,500–7,500	20–30 wt%	1,000–2,400	Two rectangular heat exchangers	(Ionescu et al. 2007)
Water-based Al_2O_3 nanofluids	Re = 600–700	0.01–0.3 vol.%	1,500–1,700	Circular tube	(Hwang et al. 2009)
TBAB hydrate slurry	Laminar flow	0–26 wt.%	1,500–4,000	Loop with storage tank and horizontal tube	(Song et al. 2009)
TBAB hydrate slurry	Re_{MR} = 1,390–2,064	5.8–17.5 vol.%	1,500–3,700	Tank with heating tube	(Ma et al. 2010)
CO_2 hydrate slurry	Re = 100–2,300	0–14 vol.%	2,000–3,500	Pilot loop with storage tank and heating tube	(Oignet et al. 2017b)

Table 7.4. *Convective heat transfer coefficient (CHTC) for various slurries and various experimental apparatus*

7.4.4. *Kinetic criterion*

As a crystallization process, hydrate formation is monitored by various successive stages, such as nucleation (first crystal appearance), growth and potentially agglomeration. Each of these steps plays an important role in the hydrate-based process. The number of particles, but also their size distribution, their morphology and their evolution over time, can indeed change many essential properties involved in refrigeration, such as flow, heat transfer and mass transfer.

In the case of hydrate-based refrigeration applications, the objective is to control hydrate formation rate while preventing agglomeration within the slurry. This issue related to agglomeration is similar to that of oil and gas applications (but in aqueous solutions for refrigeration), while the need to control hydrate formation is identical to that of gas capture and storage applications.

Several authors tried to determine hydrate formation kinetics and associated limiting factors. The first attempt to model hydrate formation kinetics (Vysniauskas and Bishnoi 1983, 1985) showed that CH_4 hydrate formation was a function of water–gas interfacial area and supercooling, and that the induction time (i.e. time before nucleation) decreases with decreasing temperature. Induction time can also be decreased by increasing stirring action (Englezos et al. 1987a). These authors proposed the first model taking accounts for hydrate growth as a crystallization process, in which gas consumption rate is linked to crystal size distribution (CSD). They described hydrate growth rate in two consecutive steps: dissolved-gas-in-liquid diffusion around hydrate crystals (liquid–solid interface); and then gas incorporation in hydrate. Later, a simplified model with assumption on mass transfer resistance (Skovborg and Rasmussen 1994) was proposed. As previously on CH_4 hydrates (Englezos et al. 1987a), a more recent work (Clarke and Bishnoi 2005) was devoted to the effects of stirring velocity on CO_2 hydrate formation and showed that CO_2 consumption rate increases with stirring velocity.

Several techniques have been used to experimentally investigate gas hydrate formation, among which FBRM (focused beam reflectance measurement) was found to be a suitable method to characterize hydrates' nucleation, growth and agglomeration (Devarakonda et al. 1999; Clarke and Bishnoi 2005; Greaves et al. 2008). FBRM is an in situ counting and sizing technique capable of measuring particle chord length ranging from 1 to 1,000 μm. It can be used to detect CO_2 hydrate nucleation by a sharp increase in the total number of particles (Clarke and Bishnoi 2005), corresponding to a sudden rise of temperature (exothermic phenomenon). Then, hydrate particle growth can be observed as an increase in the number of large particles, and agglomeration corresponds to a sudden drop in the total number of particles. FBRM data can be used to determine kinetic parameters such as hydrate growth rate, based on population balance equation (PBE) (Hulburt and Katz 1964; Randolph and Larson 1988). Such a modeling approach has been already applied to gas hydrates (Englezos et al. 1987b, 1987a).

A FTIR (Fourier transformed infrared) spectrometry analyzer connected to an in situ ATR (attenuated total reflection) probe was also implemented to determine the concentration of CO_2 dissolved in the liquid phase during CO_2 hydrate formation (Boufares et al. 2018). This work showed that hydrates consumed dissolved CO_2 much quicker than the CO_2 diffusion from vapor to liquid phase, leading to a nearly instantaneously equilibrium between liquid and hydrate phases.

Various studies on gas treatment management have focused on methods to increase hydrate formation rate and enlarge gas capacity, as kinetic additive (surfactant) use, stirring, bubbling and formation in porous media. Regarding the study of kinetic additives in natural gas storage applications, SDS is the most used as hydrate formation promoter (Lin et al. 2004; Ganji et al. 2007a, 2007b; Okutani et al. 2008). Many studies in the field of refrigeration were also dedicated to kinetic studies of hydrates, including hydrates of refrigerant gas (Bi et al. 2004, 2006, 2009, 2012; Li et al. 2004, 2005, 2006, 2008; Xie et al. 2005a, 2005c; Hashemi et al. 2015; Zafara et al. 2015), CO_2 (Jerbi et al. 2010a; Egenolf-Jonkmanns et al. 2012; Fischer et al. 2014; Xie et al. 2014; Sun and Kang 2015), quaternary salt such as TBAB (Sakamoto et al. 2011; Clain et al. 2015; Mahmoudi et al. 2016; Wang et al. 2016), HCs (Nakajima et al. 2008; Xie et al. 2009) and THF (Xie et al. 2008; Wang et al. 2012; Sun and Kang 2015; Mahmoudi et al. 2016).

In cold storage processes, the objective is to accelerate or control the charging and discharging steps. However, according to the considered step, the temperature difference between sources (hot or cold) and hydrates may not be the same. Thus, the hydrate storage process can be limited by either mass or heat transfers (Bi et al. 2009). It is therefore important to choose the suitable additive or intensification process to implement. Consequently, as for gas treatment applications, various additives were used in refrigeration to promote and accelerate hydrate formation (nucleation or growth), such as SDS (Li et al. 2004, 2005; Watanabe et al. 2005; Xie et al. 2005a, 2009, 2010; Okutani et al. 2007; Choi et al. 2014; Hashemi et al. 2015; Wang et al. 2015a), nanoparticles or metallic compounds (Li et al. 2005, 2006, 2008; Choi et al. 2014; Wang et al. 2015a; Zafara et al. 2015), zeolites (Wang et al. 2012), hydrogel additives of different morphological structure (Fischer et al. 2014) and other additives (Bi et al. 2006, 2009; Egenolf-Jonkmanns et al. 2012).

Recently, several studies were dedicated to hydrate-based processes using porous media. In fact, in typical cold thermal energy systems (CTES), hydrates are stored in the form of slurry within a tank or a loop (Yang et al. 2008; Jerbi et al. 2010a; Wang et al. 2017). The use of slurry improves heat transfer in flow (Oignet et al. 2017b) but is less suitable for static storage due to low-energy density resulting from limited mass transfer at the gas–liquid interface (Boufares et al. 2018).

According to a recent review of literature (Wang et al. 2020), enhancing mass transfer by forming CO_2 hydrates in porous media could improve CTES design, but related current research is focused on other applications such as gas separation (Yang et al. 2015), CO_2 capture (Yang et al. 2013) or sequestration (Oya et al. 2017), and mainly on CH_4 hydrates (Zhang et al. 2017). Currently, one research group is interested in cold storage by CO_2 hydrates in porous media (Benmesbah et al. 2020; Benmesbah 2021).

7.4.5. *Energy and environment criterion*

In order to be implemented in refrigeration processes, hydrate systems must be able to store a significant amount of thermal energy, transport this energy and transfer it to users. The ability of hydrate slurry to store, carry and release energy mainly depends on hydrate phase change enthalpy, but also hydrate fraction in the medium. The product of these two quantities is representative of the energy density of the hydrate slurry. A high-energy density is essential to minimize the exergy losses related to pumping and optimize installation design, flexibility and efficiency. Thermodynamic equilibrium conditions and phase change kinetics also play a vital role on the feasibility of hydrate-based cold storage processes.

Several studies are thus based on similar methods called exergy approach, entropy generation minimization, or thermodynamic analysis to assess hydrate-based cold storage process efficiency (Bi et al. 2010, 2012, 2014; Zafara et al. 2015; Pons et al. 2018a, 2018b). COP approaches, more conventional methods in refrigeration, but related to thermodynamic analysis, were also used (Shi et al. 2013, 2014a).

Because of the growing interest of CO_2 hydrate for refrigeration applications, a first economic evaluation of CO_2 hydrate cooling systems was carried out based on life-cycle cost (LCC) analysis and showed that it could be considered as an alternative to classical district cooling systems (Choi et al. 2015).

Thermal performances of refrigerant gas hydrate-based systems using direct contact (Gadalla and Ahmed 2013) or fluidized bed (Zhou et al. 2015) avoiding the use of heat exchanger in the storage tank were also investigated.

The dynamic characteristics of various chilled water storage systems, including the gas hydrate cold storage system, were also analyzed and discussed (Fang et al. 2016).

Finally, other energy studies are related to the development of a refrigerant gas hydrate-based refrigeration system (Ogawa et al. 2006), small-scale refrigerant gas-hydrate cold storage apparatus (Xie et al. 2010), CO_2 hydrate slurry formation in semi-batch stirred tank reactor and dissociation in a circulation loop with heat exchangers (Jerbi et al. 2010a), refrigerant gas hydrate-based cold storage and release processes and its industrial applicability (Wu et al. 2012), real size air-conditioning system using TBAB hydrates (Douzet et al. 2013) or hydrate-based thermal energy storage in buildings (Castellani et al. 2014).

7.5. Hydrate applications in refrigeration and air-conditioning

Slurry systems are distinguished by generation methods and architectures. In the following, various slurry generators and various examples of hydrate-based refrigeration systems at different scales are presented.

7.5.1. *Slurry generation methods*

Ice slurries were first produced by thermal processes in the last century. Initially composed of water and slush ice, they were mainly used to cool the mines. The production of ice slurry with fine crystals (diameter lower than 1 mm) was later applied to other systems, such as display cabinets in supermarket (Kauffeld et al. 2005). Different types of generators are currently used for the production of ice slurries (scrapped or brushed surface, vacuum cooling, supercooling and direct contact).

In the case of hydrate slurries, according to the guest molecules enclosed in the structure, various crystallization processes can also be used to form hydrates. The methods of generation of salt semiclathrate hydrate slurries are similar to those of ice slurries, but they differ in the case of gas hydrate slurry, as they can be formed by gas injection. Various conventional methods for hydrate slurry generation are listed below (Zhang et al. 2010b; Delgado et al. 2012; Li et al. 2012; Youssef et al. 2013):

– Scraped or brushed surface generator: the heat exchanger consists usually of two coaxial cylinders. The secondary refrigerant circulates inside the inner cylinder while the primary refrigerant flows in the jacket formed between the two cylinders and cools the secondary refrigerant fluid. The crystals (hydrates) formed in the inner tube are removed by scraping or brushing and mixed with the liquid phase. However, the method has some disadvantages, since the scraped or brushed surface generators are limited in power and have moderate levels of reliability (Lugo et al. 2006; Daitoku and Utaka 2007; 2010).

– Direct-contact generator: primary refrigerant is injected within secondary fluid to directly exchange heat with water and form hydrates (Carbajo 1985; Mori et al. 1989a, 1991; Utaka et al. 1993) or ice (Isobe and Mori 1992) in a crystallizer (evaporator, vessel). In the case of CO_2 hydrate slurry production, CO_2 is injected in a pre-cooled aqueous solution, corresponding to thermodynamic conditions for hydrate formation (Marinhas et al. 2006, 2007). However, the issue is to manage and recover the injected gas.

7.5.2. *Examples of hydrate-based refrigeration systems*

Currently, ice slurries are the most common slurries applied to refrigeration, with several hundred plants worldwide (particularly in Japan). Based on the studies of Bellas and Tassou (2005) and Kauffeld et al. (2005), ice slurries are found in various areas: food industry; transport: plane cooling for cold transport over long distances; pharmaceutical park cooling; air conditioning and refrigeration distribution networks.

One of the largest ice slurry–based achievements applied to air conditioning systems is located in the CAPCOM building in Osaka (Kuriyama and Sawahata 2001). This system consists of a scraped surface generator and storage tanks located at the top of the tower and a distribution network for 17 floors.

As early as the 1980s, hydrate-based cold storage systems were studied with refrigerant gas hydrates. Mori et al. (1989a) studied the incorporation of a CFC-12 hydrate storage tank into a vapor compression refrigerator loop (Figure 7.7). Refrigerant gas hydrate appeared in the form of slurry particles at the bottom of the tank or solidified foam floating at the surface of the water phase.

TBAB hydrate slurries were developed for the first time in Japan and used in district cooling plants for air conditioning of office buildings (NKK Corporation) (Fukushima et al. 1999; Takao et al. 2001; Ogoshi and Takao 2004).

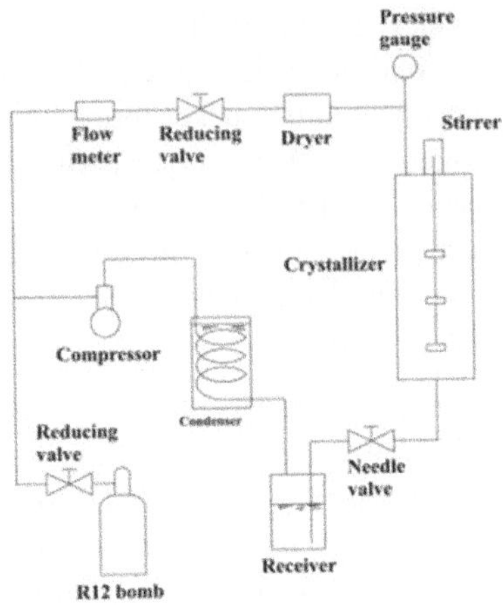

Figure 7.7. *CFC-12 hydrate-based cool storage system (Mori et al. 1989a)*

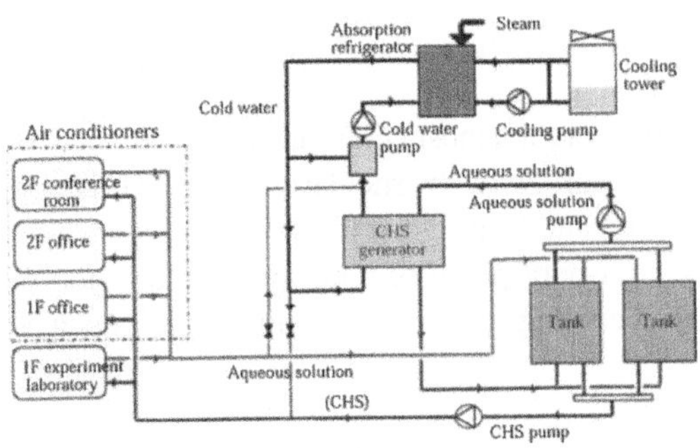

Figure 7.8. *Air-conditioning system by TBAB hydrate slurry for NKK Corporation offices in Japan (Takao et al. 2001)*

Figure 7.8 shows the system consisting of a generator, storage tanks and a distribution network of TBAB hydrate slurry. The authors worked on generator types, such as scraped surface (Takao et al. 2002) or supercooling (Takao et al. 2005). According to their results, hydrate slurry installation has a reduced size (pipes, reservoirs), which could result in a 50% reduction in overall energy consumption (reduced pumping power).

Since then, several teams were interested in salt semiclathrate or gas (refrigerant, CO_2) hydrate slurries for secondary refrigeration applications as cold storage and distribution.

Bi et al. (2004) developed a HCFC-141b hydrate cold storage system, with a cold storage tank and cold/hot reservoirs. The objective was to optimize the volumetric flow rate in the crystallizer in order to improve the formation rate of HCFC-141b hydrate (up to 0.4 $g \cdot s^{-1}$) and the cooling energy stored.

A novel hydrate based refrigeration system was also proposed (Ogawa et al. 2006). It consists of two reactors, one for hydrate formation (releasing heat to the environment) connected to a compressor for mixing and compressing gas and water, and another for hydrate dissociation (absorbing heat from the environment), a slurry system connected between the two reactors (Figure 7.9). HFC-32 + cyclopentane and water was selected as the working fluid for the simulation. Results indicated that the hydrate-based refrigeration system could be better than conventional refrigeration systems in thermodynamic efficiency. In the laboratory-scale system, HCFC-22 was used as the guest molecule, and the system could work continuously for 2 h.

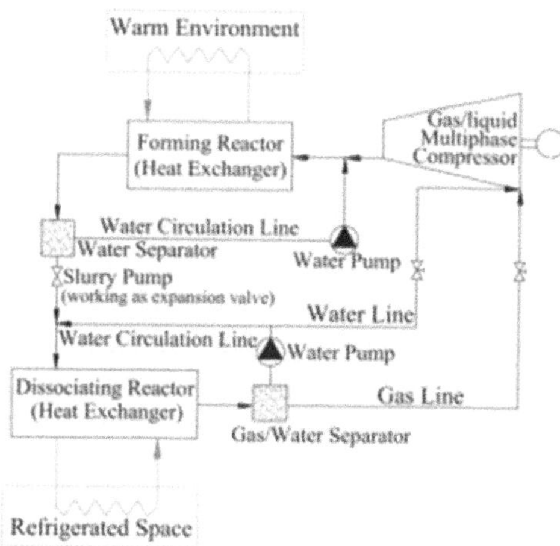

Figure 7.9. *Refrigerant hydrate-based refrigeration system (Ogawa et al. 2006)*

Another research team (Jerbi et al. 2010b) presented a refrigeration loop where CO_2 hydrate slurry was used as a two-phase secondary fluid (Figure 7.10). The system was composed of two parts, one for CO_2 hydrate slurry formation and storage (tank reactor) and another for its circulation and dissociation (loop and heat exchanger). The tank reactor had a capacity of 25 L and provided a heat flux of 1–5 kW for a hydrate fraction between 10 and 30 vol%. When the CO_2 hydrate slurry flowed in the loop, the heat exchanger allows its dissociation.

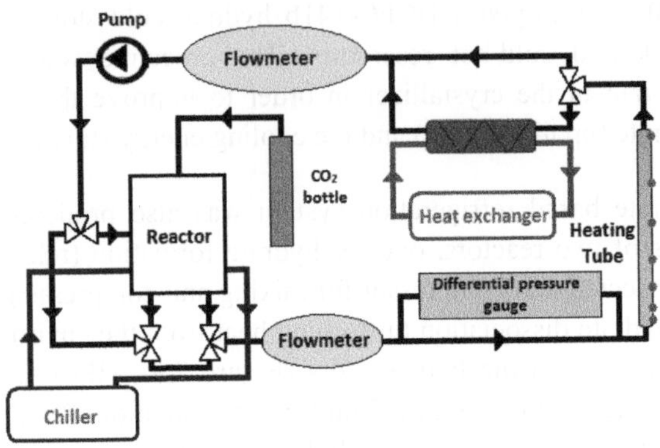

Figure 7.10. *CO_2 hydrate-based refrigeration loop (Jerbi et al. 2010b; Dufour et al. 2017, 2019). For a color version of this figure, see www.iste.co.uk/odru/heat1.zip*

Based on the device previously developed (Jerbi et al. 2010b), the impact of operating conditions (pressure, temperature) on the dynamic behavior of CO_2 hydrate slurry was studied in a stirred tank reactor for cold thermal energy storage (Dufour et al. 2017) and in coil heat exchanger for cold thermal energy restitution (Dufour et al. 2019).

In addition, since large amounts of CO_2 are involved in CO_2 hydrate formation and dissociation, a modeling approach was developed (Youssef et al. 2019) to manage vapor release in CO_2-hydrate-based refrigeration processes (Figure 7.11).

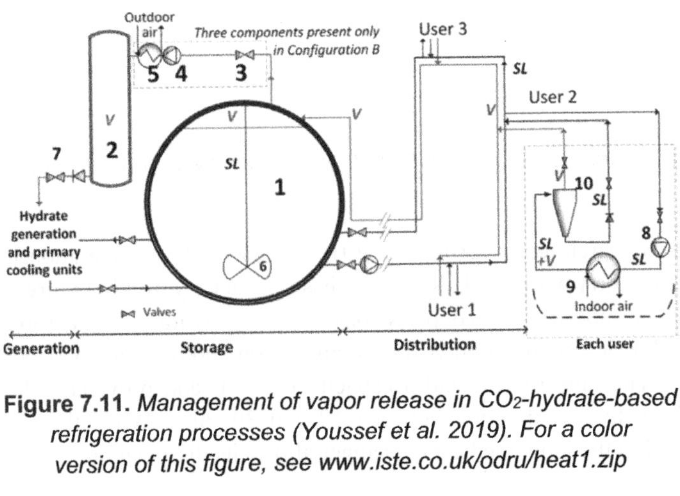

Figure 7.11. *Management of vapor release in CO_2-hydrate-based refrigeration processes (Youssef et al. 2019). For a color version of this figure, see www.iste.co.uk/odru/heat1.zip*

More recently, a novel two-stage CO_2 hydrate-based vapor-compression refrigeration system was modeled (Hua et al. 2022).

The proposed system applied pure CO_2 as the primary refrigerant and arranged both hydrate formation and dissociation at the low-pressure stage (Figure 7.12).

They developed thermodynamic and economic models to evaluate the performances of the proposed system using load-leveling storage and full storage operations and compared them with those of a conventional CO_2 single-stage vapor-compression refrigeration system. They showed that using full storage is the most efficient system.

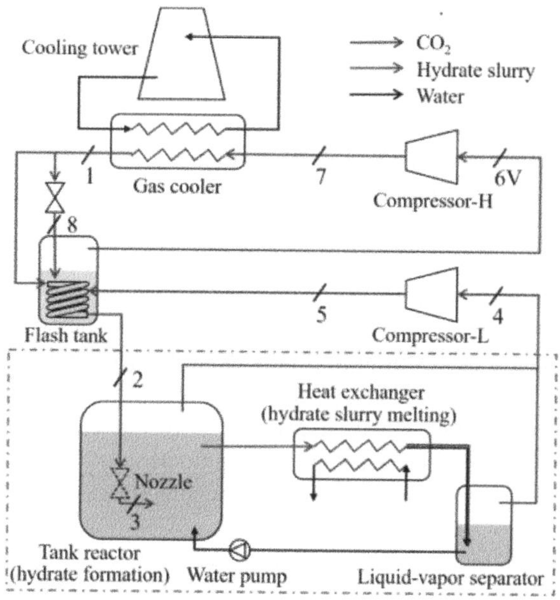

Figure 7.12. *Carbon dioxide hydrate-based vapor-compression refrigeration system (Hua et al. 2022). For a color version of this figure, see www.iste.co.uk/odru/heat1.zip*

7.6. Conclusion

Currently, the thermodynamic properties of a large number of hydrates have been known for long time in the case of gas hydrates, and more recently for single salt hydrates and mixed gas + salt hydrates. Hydrates have equilibrium properties and high dissociation enthalpy to store large amounts of energy in suitable thermodynamic conditions. In addition, various studies are focused on rheological and thermal characterization of hydrate slurries in different aqueous or organic media. Hydrate slurries are theoretically able to carry and transfer the cold with suitable flowing conditions and efficient heat exchange requirements.

Thus, a number of secondary refrigeration achievements using slurries have energy performance close to direct expansion systems, when optimizing the choice of secondary refrigerants and facility management, and combining energy storage. Nevertheless, to make hydrate-slurry-based refrigeration technology economically viable, it is estimated that the investment cost of slurry systems (generator, storage tank, distribution network) should be reduced about 20–30%. In order to disseminate hydrate slurry technology, innovation on new processes of high-efficiency slurry generation would be a great challenge.

The next scientific and technological barrier for industrial application is the control of hydrate crystallization/melting kinetics and CSD. It represents an important step to be considered in the definition, design and management of high-efficiency hydrate-based refrigeration systems. Kinetic properties are indeed crucial for the reliability and efficiency of the cooling process, since they are related to mass and heat transfers during hydrate slurry formation, flow and dissociation. As a gas-consuming phenomenon, hydrate crystallization and melting are limited by gas transfers and reactions within the slurry. Moreover, hydrate particles in slurries can agglomerate, increasing fluid viscosity and consequently pumping power in the refrigeration loop, and can even form a plug as in the case of oil and gas industry systems. Despite all the methods of improvement, prevention and remediation (chemical – thermodynamic, kinetic, dispersive – hydraulic, thermal, mechanical), hydrate formation/dissociation kinetics and agglomeration remain a concern in thermal energy management applications, and ongoing research is being conducted in this field.

7.7. References

Abhat, A. (1983). Low temperature latent heat thermal energy storage: Heat storage materials. *Solar Energy*, 30(4), 313–332.

Adisasmito, S., Franck, R.J., Sloan, E.D. (1991). Hydrates of carbon dioxide and methane mixtures. *Journal of Chemical & Engineering Data*, 36, 68–71.

Ahmed, S. and Gadalla, M. (2013). An experimental investigation of a cold thermal energy storage system. In *Advances in Energy Science and Technology*, Tang, X. Chen, Y., Dong, X., Wei, Yang, Q. (eds). International Conference on Sustainable Energy and Environmental Engineering (ICSEEE 2012), Guangzhou, December 29–30. Trans Tech Publications Ltd.

Akiya, T., Owa, M., Nakaiwa, M., Kawasaki, S., Ninemoto, M., Ando, Y. (1987). Cool storage using gas hydrate. In *17th International Congress of Refrigeration*, Vienna, 19870824–19870829.

Akiya, T., Oowa, M., Nakaiwa, M., Kawasaki, S. (1988). Crystallization characteristics of flon gas hydrates used as cool storage materials. *Kagaku Kogaku Ronbunshu*, 14(5), 692–695.

Akiya, T., Shimazaki, T., Oowa, M., Nakaiwa, M., Nakane, T., Hakuta, T. (1997a). Phase equilibria of some alternative refrigerants hydrates and their mixtures using for cool storage materials. In *32nd Intersociety Energy Conversion Engineering Conference*, Honolulu, July 27–August 2.

Akiya, T., Shimazaki, T., Oowa, M., Nakaiwa, M., Nakane, T., Usuda, T., Ebinuma, T., Kamesaki, K. (1997b). Formation characteristics of tetrahydrofuran hydrate to be used as a cool storage medium. In *62nd Annual Meeting of the Society of Chemical Engineers*, Tokyo.

Akiya, T., Shimazaki, T., Oowa, M., Matsuo, M., Yoshida, Y. (1999). Formation conditions of clathrates between HFC alternative refrigerants and water. *International Journal of Thermophysics*, 20(6), 1753–1763.

Andersson, V. and Gudmundsson, J.S. (1999). Flow experiments on concentrated hydrate slurries. In *SPE Annual Technical Conference and Exhibition: Production Operations and Engineering – General*, Houston.

Andersson, V. and Gudmundsson, J.S. (2000). Flow properties of hydrate-in-water slurries. *Annals of the New York Academy of Sciences*, 912, 322–329.

Arjmandi, M., Chapoy, A., Tohidi, B. (2007). Equilibrium data of hydrogen, methane, nitrogen, carbon dioxide, and natural gas in semi-clathrate hydrates of tetrabutyl ammonium bromide. *Journal of Chemical & Engineering Data*, 52(6), 2153–2158.

Austvik, T. and Bjorn, M. (1992). *The Transportability of Hydrates in Pipes*. Multiphase Transportation III, Roros.

Austvik, T., Li, X., Gjertsen, L.H. (2000). Hydrate plug properties. *Annals of the New York Academy of Sciences*, 973, 294–303.

Ayel, V., Lottin, O., Peerhossaini, H. (2003). Rheology, flow behaviour and heat transfer of ice slurries: A review of the state of the art. *International Journal of Refrigeration*, 26(1), 95–107.

Babu, P., Paricaud, P., Linga, P. (2016). Experimental measurements and modeling of the dissociation conditions of semiclathrate hydrates of tetrabutyl ammonium nitrate and carbon dioxide. *Fluid Phase Equilibria*, 413, 80–85.

Bellas, I. and Tassou, S.A. (2005). Present and future applications of ice slurries. *International Journal of Refrigeration*, 28(1), 115–121.

Benmesbah, F.D. (2021). Kinetic and thermodynamic study of gas hydrates in porous media: Application to gas hydrates-bearing sediments and cold storage process. PhD Thesis, UPMC.

Benmesbah, F.D., Ruffine, L., Clain, P., Osswald, V., Fandino, O., Fournaison, L., Delahaye, A. (2020). Methane hydrate formation and dissociation in sand media: Effect of water saturation, gas flowrate and particle size. *Energies*, 13(19), 5200.

Bi, Y., Guo, T., Zhu, T., Fan, S., Liang, D., Zhang, L. (2004). Influence of volumetric-flow rate in the crystallizer on the gas-hydrate cool-storage process in a new gas-hydrate cool-storage system. *Applied Energy*, 78(1), 111–121.

Bi, Y., Guo, T., Zhu, T., Zhang, L., Chen, L. (2006). Influences of additives on the gas hydrate cool storage process in a new gas hydrate cool storage system. *Energy Conversion and Management*, 47(18/19), 2974–2982.

Bi, Y., Guo, T., Zhang, L., Zhang, H., Chen, L. (2009). Experimental study on cool release process of gas-hydrate with additives. *Energy and Buildings*, 41(1), 120–124.

Bi, Y., Guo, T., Zhang, L., Chen, L., Sun, F. (2010). Entropy generation minimization for charging and discharging processes in a gas-hydrate cool storage system. *Applied Energy*, 87(4), 1149–1157.

Bi, Y., Chen, L., Sun, F. (2012). Thermodynamic optimization for crystallization process of gas hydrate. *International Journal of Energy Research*, 36(2), 269–276.

Bi, Y., Liu, X., Jiang, M. (2014). Exergy analysis of a gas-hydrate cool storage system. *Energy*, 73, 908–915.

Bo, H., Gustafsson, E.M., Setterwall, F. (1999). Tetradecane and hexadecane binary mixtures as phase change materials (PCMs) for cool storage in district cooling systems. *Energy*, 24(12), 1015–1028.

Boufares, A., Provost, E., Dalmazzone, D., Osswald, V., Clain, P., Delahaye, A., Fournaison, L. (2018). Kinetic study of CO_2 hydrates crystallization: Characterization using FTIR/ATR spectroscopy and contribution modeling of equilibrium/non-equilibrium phase-behavior. *Chemical Engineering Science*, 192, 371–379.

Camargo, R. and Palermo, T. (2002). Rheological properties of hydrate suspensions in an asphaltenic crude oil. In *4th International Conference on Gas Hydrates*, Yokohama.

Camargo, R., Palermo, T., Sinquin, A., Glenat, P. (2000). Rheological characterization of hydrate suspensions in oil dominated systems. *Annals of the New York Academy of Sciences*, 912, 906–916.

Cameirao, A., Fezoua, A., Ouabbas, Y., Herri, J.M., Darbouret, M., Sinquin, A., Glénat, P. (2011). Agglomeration of gas hydrate in a water-in-oil emulsion: Experimental and modeling studies. In *7th International Conference on Gas Hydrates*, Edinburgh.

Carbajo, J.J. (1985). A direct-contact-charged direct-contact-discharged cool storage system using gas hydrate. *ASHRAE Trans.*, 91(2A), 258–266.

Castellani, B., Morini, E., Filipponi, M., Nicolini, A., Palombo, M., Cotana, F., Rossi, F. (2014). Clathrate hydrates for thermal energy storage in buildings: Overview of proper hydrate-forming compounds. *Sustainability*, 6(10), 6815–6829.

Chami, N., Bendjenni, S., Dalmazzone, D., Clain, P., Osswald, V., Delahaye, A., Fournaison, L. (2021). Thermodynamic characterization of mixed gas hydrates in the presence of cyclopentane as guest molecule for an application in secondary refrigeration. *Chemical Engineering Science*, 244, 116790.

Chapoy, A., Anderson, R., Tohidi, B. (2007). Low-pressure molecular hydrogen storage in semi-clathrate hydrates of quaternary ammonium compounds. *Journal of the American Chemical Society*, 129(4), 746–747.

Chatti, I., Delahaye, A., Fournaison, L., Petitet, J.P. (2005). Benefits and drawbacks of clathrate hydrates: A review of their areas of interest. *Energy Conversion and Management*, 46(9/10), 1333–1343.

Choi, J.W., Chung, J.T., Kang, Y.T. (2014). CO_2 hydrate formation at atmospheric pressure using high efficiency absorbent and surfactants. *Energy*, 78, 869–876.

Choi, J.W., Kim, S., Kang, Y.T. (2015). CO_2 hydrate cooling system and LCC analysis for energy transportation application. *Applied Thermal Engineering*, 91, 11–18.

Clain, P., Delahaye, A., Fournaison, L., Mayoufi, N., Dalmazzone, D., Fürst, W. (2012). Rheological properties of tetra-n-butylphosphonium bromide hydrate slurry flow. *Chemical Engineering Journal*, 193/194, 112–122.

Clain, P., Ndoye, F.T., Delahaye, A., Fournaison, L., Lin, W., Dalmazzone, D. (2015). Particle size distribution of TBPB hydrates by focused beam reflectance measurement (FBRM) for secondary refrigeration application. *International Journal of Refrigeration*, 50, 19–31.

Clarke, M.A. and Bishnoi, P.R. (2005). Determination of the intrinsic kinetics of CO_2 gas hydrate formation using in situ particle size analysis. *Chemical Engineering Science*, 60(3), 695–709.

Claussen, W.F. (1951). A second water structure for inert gas hydrates. *J. Chem. Phys.*, 19, 1425–1426.

Daitoku, T. and Utaka, Y. (2007). An effect of scraper shapes on detachment of solid adhered to cooling surface for formation of clathrate hydrate slurry. *Heat Transfer – Asian Research*, 36(8), 489–500.

Daitoku, T. and Utaka, Y. (2010). Separation characteristics of clathrate hydrates from a cooling plate for efficient cold energy storage. *Applied Energy*, 87(8), 2682–2689.

Dalmazzone, D., Sales Silva, L.P., Delahaye, A., Fournaison, L. (2017). Calorimetric characterization of clathrate and semiclathrate hydrates. In *Gas Hydrates 1*, Broseta, D., Ruffine, L., Desmedt, A. (eds). ISTE Ltd, London, and John Wiley & Sons, New York.

Darbouret, M., Cournil, M., Herri, J.-M. (2005). Rheological study of TBAB hydrate slurries as secondary two-phase refrigerants. *International Journal of Refrigeration*, 28(5), 663–671.

Davidson, D.W. (1973). Clathrate hydrates. In *Water: A Comprehensive Treatise. Volume 2*, Franks, F. (ed.), Plenum Press, New York.

Davy, H. (1811a). The Bakerian lecture: On some of the combinations of oxymuriatic gas and oxygene, and on the chemical relations of these principles, to inflammable bodies. *Philosophical Transactions of the Royal Society of London*, 101, 1–35.

Davy, H. (1811b). On a combination of oxymuriatic gas and oxygene gas. *Philosophical Transactions of the Royal Society of London*, 101, 155–162.

Deaton, W.M. and Frost, E.M.J. (1946). Gas hydrates and their relation to the operation of natural gas pipelines. Technical Report, Bureau of Mines, Amarillo, TX.

Delahaye, A., Fournaison, L., Marinhas, S., Chatti, I., Petitet, J.P., Dalmazzone, D., Furst, W. (2006). Effect of THF on equilibrium pressure and dissociation enthalpy of CO_2 hydrates applied to secondary refrigeration. *Industrial & Engineering Chemistry Research*, 45(1), 391–397.

Delahaye, A., Fournaison, L., Marinhas, S., Martínez, M.C. (2008). Rheological study of CO_2 hydrate slurry in a dynamic loop applied to secondary refrigeration. *Chemical Engineering Science*, 63(13), 3551–3559.

Delahaye, A., Fournaison, L., Jerbi, S., Mayoufi, N. (2011). Rheological properties of CO_2 hydrate slurry flow in the presence of additives. *Industrial & Engineering Chemistry Research*, 50(13), 8344–8353.

Delahaye, A., Fournaison, L., Dalmazzone, D. (2018). Use of hydrates for cold storage and distribution in refrigeration and air-conditionning applications. In *Gas Hydrates 2*, Ruffine, L., Broseta, D., Desmedt, A. (eds). ISTE Ltd, London, and John Wiley & Sons, New York.

Delgado, M., Lazaro, A., Mazo, J., Zalba, B. (2012). Review on phase change material emulsions and microencapsulated phase change material slurries: Materials, heat transfer studies and applications. *Renewable and Sustainable Energy Reviews*, 16(1), 253–273.

Devarakonda, S., Groysman, A., Myerson, A.S. (1999). THF-water hydrate crystallization: An experimental investigation. *Journal of Crystal Growth*, 204(4), 525–538.

Douzet, J., Kwaterski, M., Lallemand, A., Chauuy, F., Flick, D., Herri, J.M. (2013). Prototyping of a real size air-conditioning system using a tetra-n-butylammonium bromide semiclathrate hydrate slurry as secondary two-phase refrigerant – Experimental investigations and modelling. *International Journal of Refrigeration*, 36(6), 1616–1631.

Dufour, T., Hoang, H.M., Oignet, J., Osswald, V., Clain, P., Fournaison, L., Delahaye, A. (2017). Impact of pressure on the dynamic behavior of CO_2 hydrate slurry in a stirred tank reactor applied to cold thermal energy storage. *Applied Energy*, 204(supplement C), 641–652.

Dufour, T., Hoang, H.M., Oignet, J., Osswald, V., Fournaison, L., Delahaye, A. (2019). Experimental and modelling study of energy efficiency of CO_2 hydrate slurry in a coil heat exchanger. *Applied Energy*, 242, 492–505.

Dyadin, Y.A. and Udachin, K.A. (1984). Clathrate formation in water-peralkylonium salts systems. *Journal of Inclusion Phenomena and Molecular Recognition in Chemistry*, 2(1/2), 61–72.

Dyadin, Y.A., Kuznetsov, P.N., Yakovlev, I.I., Pyrinova, A.V. (1973). The system water-tetrahydrofuran in the crystallization region at pressures up to 9 kbar. *Dokl. Chem.*, 208(9), 103–106.

Egenolf-Jonkmanns, B., Bruzzano, S., Deerberg, G., Fischer, M., Marzi, T., Tyukavina, M., Gomez, J.S., Wack, H., Zeidler-Fandrich, B. (2012). Low temperature chemical reaction systems for thermal storage. In *1st International Conference on Solar Heating and Cooling for Buildings and Industry*, 9–11 July, San Francisco.

Englezos, P. (1993). Clathrate hydrates. *Industrial & Engineering Chemistry Research*, 32(7), 1251–1274.

Englezos, P., Kalogerakis, N., Dholabhai, P.D., Bishnoi, P.R. (1987a). Kinetics of formation of methane and ethane gas hydrates. *Chemical Engineering Science*, 42(11), 2647–2658.

Englezos, P., Kalogerakis, N., Dholabhai, P.D., Bishnoi, P.R. (1987b). Kinetics of gas hydrate formation from mixtures of methane and ethane. *Chemical Engineering Science*, 42(11), 2659–2666.

European Parliament, Council of the European Union (2015). Regulation (EU) No 517/2014 of the European Parliament and of the Council of 16 April 2014 on fluorinated greenhouse gases and repealing Regulation (EC) No 842/2006. *Official Journal of the European Union*, 195–230.

Fang, G., Tang, F., Cao, L. (2016). Dynamic characteristics of cool thermal energy storage systems–A review. *International Journal of Green Energy*, 13(1), 1–13.

Faraday, M. (1823). On hydrate of chlorine. *Quarterly Journal of Science, Literature, and the Arts*, 15(29), 71–74.

Fidel-Dufour, A. and Herri, J.-M. (2002). Formation and transportation of methane hydrate slurries in a flow loop reactor: Influence of a dispersant. In *International Conference on Multiphase Flow*, Alba.

Fidel-Dufour, A., Gruy, F., Herri, J.-M. (2006). Rheology of methane hydrate slurries during their crystallization in a water in dodecane emulsion under flowing. *Chemical Engineering Science*, 61(2), 505–515.

Fischer, M., Bruzzano, S., Egenolf-Jonkmanns, B., Zeidler-Fandrich, B., Wack, H., Deerberg, G. (2014). Thermal storage by thermoreversible chemical reaction systems. In *Proceedings of the 2nd International Conference on Solar Heating and Cooling for Buildings and Industry*, vol I and II, 116–119.

Fournaison, L., Delahaye, A., Chatti, I., Petitet, J.P. (2004). CO_2 hydrates in refrigeration processes. *Industrial & Engineering Chemistry Research*, 43(20), 6521–6526.

Fowler, D.L., Loebenstein, W.V., Pall, D.B., Kraus, C.A. (1940). Some unusual hydrates of quaternary ammonium salts. *Journal of the American Chemical Society*, 62(5), 1140–1142.

Fu, L., Wang, H.X., Gao, F.Y. (2001). Phase equilibrium calculation of refrigerant and refrigerant mixture gas hydrates. *Energy Conversion and Application*, 116–119.

Fukushima, S., Takao, S., Ogoshi, H., Ida, H., Matsumoto, S., Akiyama, T., Otsuka, T. (1999). Development of high-density cold latent heat with clathrate hydrate. NKK Technical Report 166.

Gadalla, M. and Ahmed, S. (2013). Performance evaluation of a thermal energy storage. In *Advances in Energy Science and Technology*, Tang, X. Chen, Y., Dong, X., Wei, Yang, Q. (eds). Trans Tech Publications Ltd, Guangzhou.

Gainville, M., Sinquin, A., Darbouret, M. (2011). Hydrate slurry characterisation for laminar and turbulent lows in pipelines. In *7th International Conference on Gas Hydrates*, Edinburgh.

Ganji, H., Manteghian, M., Rahimi Mofrad, H. (2007a). Effect of mixed compounds on methane hydrate formation and dissociation rates and storage capacity. *Fuel Processing Technology*, 88(9), 891–895.

Ganji, H., Manteghian, M., Sadaghiani Zadeh, K., Omidkhah, M.R., Rahimi Mofrad, H. (2007b). Effect of different surfactants on methane hydrate formation rate, stability and storage capacity. *Fuel*, 86(3), 434–441.

Greaves, D., Boxall, J., Mulligan, J., Montesi, A., Creek, J., Sloan, E.D., Koh, C.A. (2008). Measuring the particle size of a known distribution using the focused beam reflectance measurement technique. *Chemical Engineering Science*, 63(22), 5410–5419.

Hammerschmidt, E.G. (1934). Formation of gas hydrates in natural gas transmission lines. *Industrial & Engineering Chemistry*, 26(8), 851–855.

Hanley, H.J.M., Meyers, G.J., White, J.W., Sloan, E.D. (1989). The melting curve of tetrahydrofuran hydrate in D_2O. *International Journal of Thermophysics*, 10(4), 903–909.

Hashemi, H., Babaee, S., Mohammadi, A.H., Naidoo, P., Ramjugernath, D. (2015). Experimental study and modeling of the kinetics of refrigerant hydrate formation. *Journal of Chemical Thermodynamics*, 82, 47–52.

Hashimoto, S., Sugahara, T., Sato, H., Ohgaki, K. (2007). Thermodynamic stability of H2 + tetrahydrofuran mixed gas hydrate in nonstoichiometric aqueous solutions. *Journal of Chemical & Engineering Data*, 52(2), 517–520.

Hashimoto, S., Sugahara, T., Moritoki, M., Sato, H., Ohgaki, K. (2008). Thermodynamic stability of hydrogen + tetra-n-butyl ammonium bromide mixed gas hydrate in nonstoichiometric aqueous solutions. *Chemical Engineering Science*, 63(4), 1092–1097.

Hashimoto, S., Kawamura, K., Ito, H., Nobeoka, M., Ohgaki, K., Inoue, Y. (2011). Rheological study on tetra-n-butyl ammonium salt semi-clathrate hydrate slurries. In *7th International Conference on Gas Hydrates*, Edinburgh.

Hayashi, K., Takao, S., Ogoshi, H., Matsumoto, S. (2000). Research and development on high-density cold latent-heat medium transportation technology. In *Fifth Workshop of IEA*, Annexe 10.

Hua, N., Lu, T., Yang, L., McKeown, A., Yu, Z., Xu, B., Sciacovelli, A., Ding, Y., Li, Y. (2022). Thermodynamic analysis and economic assessment of a carbon dioxide hydrate-based vapor compression refrigeration system using load shifting controls in summer. *Energy Conversion and Management*, 251, 114901.

Huang, L., Doetsch, C., Pollerberg, C. (2010). Low temperature paraffin phase change emulsions. *International Journal of Refrigeration*, 33(8), 1583–1589.

Hulburt, H.M. and Katz, S. (1964). Some problems in particle technology: A statistical mechanical formulation. *Chemical Engineering Science*, 19(8), 555–574.

Hwang, K.S., Jang, S.P., Choi, S.U.S. (2009). Flow and convective heat transfer characteristics of water-based Al_2O_3 nanofluids in fully developed laminar flow regime. *International Journal of Heat and Mass Transfer*, 52(1/2), 193–199.

IIF-IIR (2019). Le rôle du froid dans l'économie mondiale. 38e note d'information sur les technologies du froid, Institut international du froid – International Institute of Refrigeration.

Ionescu, C., Haberschill, P., Kiss, I., Lallemand, A. (2007). Local and global heat transfer coefficients of a stabilised ice slurry in laminar and transitional flows. *International Journal of Refrigeration*, 30(6), 970–977.

Isobe, F. and Mori, Y.H. (1992). Formation of gas hydrate or ice by direct-contact evaporation of CFC alternatives. *International Journal of Refrigeration*, 15(3), 137–142.

Jeffrey, G.A. (1984). Hydrate inclusion compounds. *Journal of Inclusion Phenomena and Macrocyclic Chemistry*, 1(3), 211–222.

Jeffrey, G.A. and McMullan, R.K. (1967). *The Clathrate Hydrates*. John Wiley & Sons, New York.

Jerbi, S., Delahaye, A., Fournaison, L., Haberschill, P. (2010a). Characterization of CO_2 hydrate formation and dissociation kinetics in a flow loop. *International Journal of Refrigeration*, 33(8), 1625–1631.

Jerbi, S., Delahaye, A., Fournaison, L., Haberschill, P. (2010b). Design of a new circulation loop and heat transfer of CO_2 hydrate slurry. In *9th IIR Conference on Phase Change Materials and Slurries for Refrigeration and Air Conditioning*, Sofia.

Jerbi, S., Delahaye, A., Oignet, J., Fournaison, L., Haberschill, P. (2013). Rheological properties of CO_2 hydrate slurry produced in a stirred tank reactor and a secondary refrigeration loop. *International Journal of Refrigeration*, 36(4), 1294–1301.

Kakiuchi, H., Yabe, M., Yamazaki, M. (2003). A study of trimethylolethane hydrate as a phase change material. *Journal of Chemical Engineering of Japan*, 36(7), 788–793.

Kang, S.-P., Lee, H., Ryu, B.-J. (2001). Enthalpies of dissociation of clathrate hydrates of carbon dioxide, nitrogen, (carbon dioxide + nitrogen), and (carbon dioxide + nitrogen + tetrahydrofuran). *The Journal of Chemical Thermodynamics*, 33(5), 513–521.

Kauffeld, M., Kawaji, M., Egolf, P.W., Melinder, A., Davies, T.W. (2005). *Handbook on Ice Slurries – Fundamentals and Engineering*. IIF-IIR, Paris.

Kumano, H., Hirata, T., Kudoh, T. (2011a). Experimental study on the flow and heat transfer characteristics of a tetra-n-butyl ammonium bromide hydrate slurry (first report: Flow characteristics). *International Journal of Refrigeration*, 34(8), 1953–1962.

Kumano, H., Hirata, T., Kudoh, T. (2011b). Experimental study on the flow and heat transfer characteristics of a tetra-n-butyl ammonium bromide hydrate slurry (second report: Heat transfer characteristics). *International Journal of Refrigeration*, 34(8), 1963–1971.

Kuriyama, T. and Sawahata, Y. (2001). Slurry ice transportation and cold distribution system. In *Fourth Workshop of IIR Ice Slurry Working Party*. Osaka, 12–13 November.

Larson, S.D. (1955). Phase studies of the two-component carbon dioxide – Water system, involving the carbon dioxyde hydrates. PhD Thesis, University of Illinois, Urbana.

Li, J., Guo, K., Liang, D., Wang, R. (2004). Experiments on fast nucleation and growth of HCFC141b gas hydrate in static water columns. *International Journal of Refrigeration*, 27(8), 932–939.

Li, J., Liang, D., Guo, K., Wang, R. (2005). The influence of additives and metal rods on the nucleation and growth of gas hydrates. *Journal of Colloid and Interface Science*, 283(1), 223–230.

Li, J., Liang, D., Guo, K., Wang, R., Fan, S. (2006). Formation and dissociation of HFC134a gas hydrate in nano-copper suspension. *Energy Conversion and Management*, 47(2), 201–210.

Li, D., Du, J., Fan, S., Liang, D., Li, X., Huang, N. (2007). Clathrate dissociation conditions for methane + tetra-n-butyl ammonium bromide (TBAB) + water. *Journal of Chemical & Engineering Data*, 52(5), 1916–1918.

Li, G., Liu, D., Xie, Y. (2008). Advances of research on refrigerant gas hydrate growth property with nanotechnology. *Cryogenics and Refrigeration Proceedings*, 928–931.

Li, G., Liu, D., Xie, Y. (2010). Study on thermal properties of TBAB-THF hydrate mixture for cold storage by DSC. *Journal of Thermal Analysis and Calorimetry*, 102(2), 819–826.

Li, G., Hwang, Y., Radermacher, R. (2012). Review of cold storage materials for air conditioning application. *International Journal of Refrigeration*, 35(8), 2053–2077.

Liang, D., Guo, K., Wang, R., Fan, S. (2001). Hydrate equilibrium data of 1, 1, 1,2-tetrafluoroethane (HFC-134a), 1,1-dichloro-1-fluoroethane (HCFC-141b) and 1,1-difluoroethane (HFC-152a). *Fluid Phase Equilibria*, 187, 61–70.

Liang, D., Guo, K., Fan, S., Wang, R. (2002). Measurement of melting heat of HCFC-141b gas hydrate with DSC. *Journal of Engineering Thermophysics*, 23, 47–49.

Lin, W. (2007). *Effect of TBAB on Thermal Properties of CO_2 Hydrate*. Cemagref, Antony.

Lin, W., Chen, G.J., Sun, C.Y., Guo, X.Q., Wu, Z.K., Liang, M.Y., Chen, L.T., Yang, L.Y. (2004). Effect of surfactant on the formation and dissociation kinetic behavior of methane hydrate. *Chemical Engineering Science*, 59(21), 4449–4455.

Lin, W., Delahaye, A., Fournaison, L. (2008). Phase equilibrium and dissociation enthalpy for semi-clathrate hydrate of CO_2 + TBAB. *Fluid Phase Equilibria*, 264(1/2), 220–227.

Lin, W., Dalmazzone, D., Fürst, W., Delahaye, A., Fournaison, L., Clain, P. (2013). Accurate DSC measurement of the phase transition temperature in the TBPB-water system. *The Journal of Chemical Thermodynamics*, 61, 132–137.

Lin, W., Dalmazzone, D., Fürst, W., Delahaye, A., Fournaison, L., Clain, P. (2014). Thermodynamic properties of semiclathrate hydrates formed from the TBAB + TBPB + water and CO_2 + TBAB + TBPB + water systems. *Fluid Phase Equilibria*, 372, 63–68.

Lipkowski, J., Komarov, V.Y., Rodionova, T.V., Dyadin, Y.A., Aladko, L.S. (2002). The structure of tetrabutylammonium bromide hydrate $(C_4H_9)4NBr.21/3H_2O$. *Journal of Supramolecular Chemistry*, 2, 435–439.

Liu, Y., Guo, K., Liang, D., Fan, S. (2003). Effects of magnetic fields on HCFC-141b refrigerant gas hydrate formation. *Science in China Series B-Chemistry*, 46(4), 407–415.

Lugo, R., Fournaison, L., Guilpart, J. (2006). Ice-liquid-vapour equilibria of ammonia and ethanol aqueous solutions applied to the production of ice-slurries – Prediction and experimental results. *Chemical Engineering and Processing*, 45(1), 66–72.

Ma, Z. and Zhang, P. (2013). Modeling the heat transfer characteristics of flow melting of phase change material slurries in the circular tubes. *International Journal of Heat and Mass Transfer*, 64, 874–881.

Ma, Z., Zhang, P., Wang, R., Furui, S., Xi, G. (2010). Forced flow and convective melting heat transfer of clathrate hydrate slurry in tubes. *International Journal of Heat and Mass Transfer*, 53(19/20), 3745–3757.

Mahmoudi, B., Naeiji, P., Varaminian, F. (2016). Study of tetra-n-butylammonium bromide and tetrahydrofuran hydrate formation kinetics as a cold storage material for air conditioning system. *Journal of Molecular Liquids*, 214, 96–100.

Marinhas, S., Delahaye, A., Fournaison, L., Dalmazzone, D., Fürst, W., Petitet, J.-P. (2006). Modelling of the available latent heat of a CO_2 hydrate slurry in an experimental loop applied to secondary refrigeration. *Chemical Engineering and Processing*, 45(3), 184–192.

Marinhas, S., Delahaye, A., Fournaison, L. (2007). Solid fraction modelling for CO_2 and CO_2-THF hydrate slurries used as secondary refrigerants. *International Journal of Refrigeration*, 30(5), 758–766.

Martínez, M.C., Dalmazzone, D., Fürst, W., Delahaye, A., Fournaison, L. (2008). Ther-modynamic properties of THF + CO_2 hydrates in relation with refrigeration applications. *AIChE Journal*, 54(4), 1088–1095.

Marum (2009). Scheme of gas hydrate structure [Online]. Available at: https://www.marum.de/en/about-us/General-Geology-Marine-Geology/General-Information-on-Gas-Hydrates.html.

Matsuo, M., Funakura, M., Yoshida, Y., Akiya, T., Oowa, M. (1999). Regenerative air-conditioning system with natural refrigerant clathrates. In *20th International Congress of Refrigeration*, IIR/IIF, Sydney.

Mayoufi, N., Dalmazzone, D., Fürst, W., Delahaye, A., Fournaison, L. (2010). CO_2 enclathration in hydrates of peralkyl-(ammonium/phosphonium) salts: Stability conditions and dissociation enthalpies. *Journal of Chemical & Engineering Data*, 55(3), 1271–1275.

Mayoufi, N., Dalmazzone, D., Delahaye, A., Clain, P., Fournaison, L., Fürst, W. (2011). Experimental data on phase behavior of simple tetrabutylphosphonium bromide (TBPB) and mixed CO_2 + TBPB semiclathrate hydrates. *Journal of Chemical & Engineering Data*, 56(6), 2987–2993.

Mayoufi, N., Dalmazzone, D., Fürst, W., Elghoul, L., Seguatni, A., Delahaye, A., Fournaison, L. (2012). Phase behaviour of tri-*n*-butylmethylammonium chloride hydrates in the presence of carbon dioxide. *Journal of Thermal Analysis and Calorimetry*, 109(1), 481–486.

McMullan, R.K. and Jeffrey, G.A. (1959). Hydrates of the tetra n-butyl and tetra i-amyl quaternary ammmonium salts. *The Journal of Chemical Physics*, 31, 1231–1234.

Mori, Y.H. and Isobe, F. (1991). A model for gas hydrate formation accompanying direct-contact evaporation of refrigerant drops in water. *Int. Commun. Heat Mass Transfer*, 18, 599–608.

Mori, Y.H. and Mori, T. (1989a). Characterization of gas hydrate formation in direct-contact cool storage process. *International Journal of Refrigeration*, 12, 259–265.

Mori, Y.H. and Mori, T. (1989b). Formation of gas hydrate with CFC alternative R-134a. *AIChE Journal*, 35(7), 1227–1228.

Nakajima, M., Ohmura, R., Mori, Y.H. (2008). Clathrate hydrate formation from cyclopentane-in-water emulsions. *Industrial & Engineering Chemistry Research*, 47(22), 8933–8939.

Nakayama, H. (1987). Hydrates of organic compounds. XI: Determination of the melting point and hydration numbers of the clathrate-like hydrate of tetrabutylammonium chloride by differential scanning calorimetry. *Bulletin of the Chemical Society of Japan*, 60, 839–843.

Nygaard, H.F. (1989). Transportability of hydrates in multi-phase systems. In *Proceedings of the 4th International Conference on Multi-phase Flow*, Nice.

Ogawa, T., Ito, T., Watanabe, K., Tahara, K.-I., Hiraoka, R., Ochiai, J.-I., Ohmura, R., Mori, Y.H. (2006). Development of a novel hydrate-based refrigeration system: A preliminary overview. *Applied Thermal Engineering*, 26(17/18), 2157–2167.

Ogoshi, H. and Takao, S. (2004). Air-conditioning system using clathrate hydrate slurry. JFE Technical Report, 3, 1–5.

Oignet, J., Delahaye, A., Torré, J.-P., Dicharry, C., Hoang, H.M., Clain, P., Osswald, V., Youssef, Z., Fournaison, L. (2017a). Rheological study of CO_2 hydrate slurry in the presence of sodium dodecyl sulfate in a secondary refrigeration loop. *Chemical Engineering Science*, 158, 294–303.

Oignet, J., Hoang, H.-M., Osswald, V., Delahaye, A., Fournaison, L., Haberschill, P. (2017b). Experimental study of convective heat transfer coefficients of CO_2 hydrate slurries in a secondary refrigeration loop. *Applied Thermal Engineering*, 118, 630–637.

Okutani, K., Kuwabara, Y., Mori, Y.H. (2007). Surfactant effects on hydrate formation in an unstirred gas/liquid system: Amendments to the previous study using HFC-32 and sodium dodecyl sulfate. *Chemical Engineering Science*, 62(14), 3858–3860.

Okutani, K., Kuwabara, Y., Mori, Y.H. (2008). Surfactant effects on hydrate formation in an unstirred gas/liquid system: An experimental study using methane and sodium alkyl sulfates. *Chemical Engineering Science*, 63(1), 183–194.

Otake, K., Tsuji, T., Sato, I., Akiya, T., Sako, T., Hongo, M.U. (2000). A proposal of a new technique for the density measurement of solids. *Fluid Phase Equilibria*, 171(1/2), 175–179.

Oya, S., Aifaa, M., Ohmura, R. (2017). Formation, growth and sintering of CO_2 hydrate crystals in liquid water with continuous CO_2 supply: Implication for subsurface CO_2 sequestration. *International Journal of Greenhouse Gas Control*, 63, 386–391.

Oyama, H., Ebinuma, T., Shimada, W., Takeya, S., Nagao, J., Uchida, T., Narita, H. (2002). Viscosity increase before nucleation of CO_2 gas hydrate. In *4th Int. Conf. on Gas Hydrates*, Yokohama.

Oyama, H., Shimada, W., Ebinuma, T., Kamata, Y., Takeya, S., Uchida, T., Nagao, J., Narita, H. (2005). Phase diagram, latent heat, and specific heat of TBAB semiclathrate hydrate crystals. *Fluid Phase Equilibria*, 234(1/2), 131–135.

Pandey, G., Linga, P., Sangwai, J.S. (2017). High pressure rheology of gas hydrate formed from multiphase systems using modified Couette rheometer. *Review of Scientific Instruments*, 88(2), 025102.

Parrish, W.R. and Prausnitz, J.M. (1972). Dissociation pressures of gas hydrates formed by gas mixtures. *Industrial & Engineering Chemistry Process Design and Development*, 11(1), 26–35.

Pauling, L. and Marsh, R.E. (1952). The structure of chlorine hydrate. *Proc. Natl. Acad. Sci. U.S.A.*, 38, 112–118.

Peng, B.-Z., Chen, J., Sun, C.-Y., Dandekar, A., Guo, S.-H., Liu, B., Mu, L., Yang, L.-Y., Li, W.-Z., Chen, G.-J. (2012). Flow characteristics and morphology of hydrate slurry formed from (natural gas plus diesel oil/condensate oil plus water) system containing anti-agglomerant. *Chemical Engineering Science*, 84, 333–344.

Peysson, Y., Maurel, P., Vilagines, R. (2003). Hydrate transportability in multiphase flow. In *11th Int. Conf. Multiphase*. BHR Group, San Remo.

Peysson, Y., Duret, E., Maurel, P., Vilagines, R. (2004). Stratified and dispersed flow of gas hydrates in offshore production. In *Hydrotransport 16th International Conference*, 26–28 April, Santiago.

Pinder, K.L. (1964). Time dependant rheology of the tetrahydrofuran-hydrogen sulphide gas hydrate slurry. *The Canadian Journal of Chemical Engineering*, 42(3), 132–138.

Pons, M., Delahaye, A., Fournaison, L., Dalmazzone, D. (2018a). Energy analysis of two-phase secondary refrigeration in steady-state operation, part 2: Exergy analysis and effects of phase change kinetics. *Energy*, 161, 1291–1299.

Pons, M., Hoang, H.M., Dufour, T., Fournaison, L., Delahaye, A. (2018b). Energy analysis of two-phase secondary refrigeration in steady-state operation, part 1: Global optimization and leading parameter. *Energy*, 161, 1282–1290.

Randolph, A. and Larson, M. (1988). *Theory of Particulate Processes: Analysis and Techniques of Continuous Crystallization*. Academic Press, New York.

Ripmeester, J.A. and Ratcliffe, C.I. (1998). The diverse nature of dodecahedral cages in clathrate hydrates as revealed by ^{129}Xe and ^{13}C NMR spectroscopy: CO_2 as a small-cage guest. *Energy & Fuels*, 12(2), 197–200.

Ripmeester, J.A., Ratcliffe, C.I., Powell, B.M. (1987). A new clathrate hydrate structure. *Nature*, 325, 135–136.

Robinson, D.B. and Mehta, B.R. (1971). Hydrates in the propane-carbon dioxide-water system. *Journal of Canadian Petroleum Technology*, 10, 33.

Rodionova, T.V., Manakov, A.Y., Stenin, Y.G., Villevald, G.V., Karpova, T.D. (2008). The heats of fusion of tetrabutylammonium fluoride ionic clathrate hydrates. *Journal of Inclusion Phenomena and Macrocyclic Chemistry*, 61(1/2), 107–111.

Rondeau-Mouro, C., Delahaye, A., Cambert, M., Fournaison, L., Mariette, F. (2015). Characterization of solid content and distinction between type A and B crystals of TBAB hydrates by time domain NMR. *Chemical Engineering Science*, 138, 544–551.

Sakamoto, H., Sato, K., Shiraiwa, K., Takeya, S., Nakajima, M., Ohmura, R. (2011). Synthesis, characterization and thermal-property measurements of ionic semi-clathrate hydrates formed with tetrabutylphosphonium chloride and tetrabutylammonium acrylate. *Rsc Advances*, 1(2), 315–322.

Sato, K., Tokutomi, H., Ohmura, R. (2013). Phase equilibrium of ionic semiclathrate hydrates formed with tetrabutylammonium bromide and tetrabutylammonium chloride. *Fluid Phase Equilibria*, 337, 115–118.

Shi, X.J. and Zhang, P. (2013). A comparative study of different methods for the generation of tetra-n-butyl ammonium bromide clathrate hydrate slurry in a cold storage air-conditioning system. *Applied Energy*, 112, 1393–1402.

Shi, X.J. and Zhang, P. (2014a). Cold storage by tetra-n-butyl ammonium bromide clathrate hydrate slurry generated with different storage approaches at 40 wt% initial aqueous solution concentration. *International Journal of Refrigeration*, 42, 77–89.

Shi, X.J. and Zhang, P. (2014b). Crystallization of tetra-n-butyl ammonium bromide clathrate hydrate slurry and the related heat transfer characteristics. *Energy Conversion and Management*, 77, 89–97.

Shi, X.J. and Zhang, P. (2016). Two-phase flow and heat transfer characteristics of tetra-n-butyl ammonium bromide clathrate hydrate slurry in horizontal 90 degrees elbow pipe and U-pipe. *International Journal of Heat and Mass Transfer*, 97, 364–378.

Shi, L., Yi, L., Shen, X., Wu, W., Liang, D. (2016). Dissociation temperatures of mixed semiclathrate hydrates formed with tetrabutylammonium bromide plus tetra-butylammonium chloride. *Journal of Chemical and Engineering Data*, 61(6), 2155–2159.

Shimada, W., Shiro, M., Kondo, H., Takeya, S., Oyama, H., Ebinuma, T., Narita, H. (2005). Tetra-n-butylammonium bromide-water (1/38). *Acta Crystallographica Section C*, 61(2), o65–o66.

Sinquin, A., Palermo, T., Peysson, Y. (2004). Rheological and flow properties of gas hydrate suspensions. *Oil & Gas Science and Technology – Rev. IFP*, 59(1), 41–57.

Skovborg, P. and Rasmussen, P. (1994). Comments on: Hydrate dissociation enthalpy and guest size. *Fluid Phase Equilibria*, 96, 223–231.

Sloan, E.D. (1998). *Clathrate Hydrates of Natural Gases*. Marcel Dekker Inc., New York.

Sloan, E.D. and Koh, C.A. (2007). *Clathrate Hydrates of Natural Gases*. CRC Press, New York.

Song, W., Xiao, R., Huang, C., He, S., Dong, K., Feng, Z. (2009). Experimental investigation on TBAB clathrate hydrate slurry flows in a horizontal tube: Forced convective heat transfer behaviors. *International Journal of Refrigeration*, 32(7), 1801–1807.

Stamatiou, E., Fournaison, L., Guilpart, J. (2005). A microscopic study of the ice slurry crystal growth process. In *Sixth Workshop on ice slurries of the IIR*, Yverdon-les-Bains.

Suginaka, T., Sakamoto, H., Iino, K., Takeya, S., Nakajima, M., Ohmura, R. (2012). Thermodynamic properties of ionic semiclathrate hydrate formed with tetrabutylphosphonium bromide. *Fluid Phase Equilibria*, 317, 25–28.

Sun, Q. and Kang, Y.T. (2015). Experimental correlation for the formation rate of CO_2 hydrate with THF (tetrahydrofuran) for cooling application. *Energy*, 91, 712–719.

Suzuki, H., Itotagawa, T., Indartono, Y.S., Usui, H., Wada, N. (2006). Rheological characteristics of trimethylolethane hydrate slurry treated with drag-reducing surfactants. *Rheologica Acta*, 46(2), 287–295.

Suzuki, H., Wada, N., Komoda, Y., Usui, H., Ujiie, S. (2009). Drag reduction characteristics of trimethylolethane hydrate slurries treated with surfactants. *International Journal of Refrigeration*, 32(5), 931–937.

Takao, S., Ogoshi, H., Matsumoto, S. (2001). New air conditioning systems using hydrate slurry. NKK Technical Report 174.

Takao, S., Ogoshi, H., Matsumoto, S. (2002). Air conditioning and thermal storage systems using clathrate hydrate slurry. United States Patent Application Publication No. US:2002/083720 A1.

Takao, S., Ogoshi, H., Matsumoto, S., Furumoto, N. (2005). Apparatus for producing hydrate slurry. United States Patent Application Publication No. US:2005/0163681 A1.

Takao, S., Ogoshi, H., Fukushima, S. (2007). Thermal storage medium using a hydrate and apparatus thereof, and method for producing the thermal storage medium. United States Patent Application Publication No. 7,246,506, B2.

Tanasawa, I. and Takao, S. (2002). Low-temperature storage using clathrate hydrate slurries of tetra-n-butylammonium bromide: Thermophysical properties and morphology of clathrate hydrate crystals. In *4th Int. Conf. on Gas Hydrates*, Yokohama.

Tanii, T., Minemoto, M., Nakazawa, K., Ando, Y. (1996a). Study on cool storage system using CFC-11 (CCl3F) clathrate. *Kagaku Kogaku Ronbunshu*, 22(2), 241–248.

Tanii, T., Minemoto, M., Nakazawa, K., Ando, Y. (1996b). Study on cool storage system using HCFC (Hydrochlorofluorocarbon)-141b (CCl2FCH3) clathrate. *Kagaku Kogaku Ronbunshu*, 22(3), 473–479.

Tanii, T., Minemoto, M., Nakazawa, K., Ando, Y. (1997). Study on a cool storage system using HCFC (hydro-chloro-fluoro-carbon)-141b(CCI2FCH3) (1,1-dichloro-1-fluoro-ethane) clathrate. *Canadian Journal of Chemical Engineering*, 75(2), 353–361.

Ternes, M.P. (1984). Characterization of refrigerant-12 gas hydrate formation for heat-pump cool storage applications. *Mechanical Engineering*, 106(5), 91.

Tse, J.S., McKinnon, W.R., Marchi, M. (1987). Thermal expansion of structure I ethylene oxide hydrate. *Journal of Physical Chemistry*, 91, 4188–4193.

Uchida, T., Takagi, A., Kawabata, J., Mae, S., Hondoh, T. (1995). Raman spectroscopic analyses of the growth process of CO_2 hydrates. *Energy Conversion and Management*, 36(6/9), 547–550.

Udachin, K.A., Ratcliffe, C.I., Ripmeester, J.A. (2001). Structure, composition, and thermal expansion of CO_2 hydrate from single crystal X-ray diffraction measurements. *The Journal of Physical Chemistry B*, 105(19), 4200–4204.

Unruh, C.H. and Katz, D.L. (1949). Gas hydrates of carbon dioxide-methane mixtures. *Trans. AIME*, 186, 83.

Utaka, Y., Saito, A., Seki, T. (1993). Gas hydrate cold-storage using direct contact heat-transfer of liquid vapor-phase change and natural circulation of refrigerant in closed vessel. *JSME International Journal Series B-Fluids and Thermal Engineering*, 36(1), 150–155.

Vlahakis, J.G., Chen, H.-S., Suwandi, M.S., Barduhn, A.J. (1972). *The Growth Rate of Ice Crystals: Properties of Carbon Dioxide Hydrate, a Review of Properties of 51 Gas Hydrates*. U.S. Department of the Interior, Syracuse.

Vysniauskas, A. and Bishnoi, P.R. (1983). A kinetic study of methane hydrate formation. *Chemical Engineering Science*, 38(7), 1061–1072.

Vysniauskas, A. and Bishnoi, P.R. (1985). Kinetics of ethane hydrate formation. *Chemical Engineering Science*, 40(2), 299–303.

Wang, X. and Dennis, M. (2015a). An experimental study on the formation behavior of single and binary hydrates of TBAB, TBAF and TBPB for cold storage air conditioning applications. *Chemical Engineering Science*, 137, 938–946.

Wang, X. and Dennis, M. (2015b). Influencing factors on the energy saving performance of battery storage and phase change cold storage in a PV cooling system. *Energy and Buildings*, 107, 84–92.

Wang, X. and Dennis, M. (2016). Characterisation of thermal properties and charging performance of semi-clathrate hydrates for cold storage applications. *Applied Energy*, 167, 59–69.

Wang, X. and Dennis, M. (2017). Charging performance of a CO_2 semi-clathrate hydrate based PCM in a lab-scale cold storage system. *Applied Thermal Engineering*, 126, 762–773.

Wang, X., Niu, J.L., Li, Y., Wang, X., Chen, B.J., Zeng, R.L., Song, Q.W., Zhang, Y.P. (2007). Flow and heat transfer behaviors of phase change material slurries in a horizontal circular tube. *International Journal of Heat and Mass Transfer*, 50(13/14), 2480–2491.

Wang, W., Fan, S., Liang, D., Yang, X. (2008). Experimental study on flow characters of CH_3CCl_2F hydrate slurry. *International Journal of Refrigeration*, 31(3), 371–378.

Wang, Y., Lang, X., Fan, S. (2012). Accelerated nucleation of tetrahydrofuran (THF) hydrate in presence of ZIF-61. *Journal of Natural Gas Chemistry*, 21(3), 299–301.

Wang, X., Dennis, M., Hou, L. (2014). Clathrate hydrate technology for cold storage in air conditioning systems. *Renewable and Sustainable Energy Reviews*, 36, 34–51.

Wang, X., Zhang, F., Lipiński, W. (2020). Carbon dioxide hydrates for cold thermal energy storage: A review. *Solar Energy*, 211, 11–30.

Watanabe, K., Imai, S., Mori, Y.H. (2005). Surfactant effects on hydrate formation in an unstirred gas/liquid system: An experimental study using HFC-32 and sodium dodecyl sulfate. *Chemical Engineering Science*, 60(17), 4846–4857.

Waycuilis, J., Sirajuddin, H., Rensing, P. (2011). A novel approach to production of hydrate slurries at high process intensity and conceptual applications. In *7th International Conference on Gas Hydrates*, Edinburgh.

Weast, R.C., Astle, M.J., Beyer, W.H. (1987). *Handbook of Chemistry and Physics*. CRC Press, Boca Raton.

Webb, E.B., Rensing, P.J., Koh, C.A., Sloan, E.D., Sum, A.K., Liberatore, M.W. (2012). High-pressure rheology of hydrate slurries formed from water-in-oil emulsions. *Energy & Fuels*, 26(6), 3504–3509.

Webb, E.B., Koh, C.A., Liberatore, M.W. (2013). Rheological properties of methane hydrate slurries formed from AOT plus water plus oil microemulsions. *Langmuir*, 29(35), 10997–11004.

Webb, E.B., Koh, C.A., Liberatore, M.W. (2014). High pressure rheology of hydrate slurries formed from water-in-mineral oil emulsions. *Industrial & Engineering Chemistry Research*, 53(17), 6998–7007.

Wu, J. and Wang, S. (2012). Research on cool storage and release characteristics of R134a gas hydrate with additive. *Energy and Buildings*, 45, 99–105.

Xiao, R., Wu, S., Tang, L., Huang, C., Feng, Z. (2006). Experimental investigation of pressure-drop of clathrate hydrate slurry (CHS) flow of tetra butyl ammonium bromide (TBAB) in straight pipe. In *Proceedings of 10th International Conference on Thermal Energy Storage*, Richard Stockton College of New Jersey.

Xie, Y., Guo, K., Liang, D., Fan, S., Gu, J. (2005a). Gas hydrate growth morphology outside of horizontal heat transfer tube. *Journal of Crystal Growth*, 276(1/2), 253–264.

Xie, Y., Guo, K., Liang, D., Fan, S., Gu, J. (2005b). Steady gas hydrate growth along vertical heat transfer tube without stirring. *Chemical Engineering Science*, 60(3), 777–786.

Xie, Y., Guo, K., Liang, D., Fan, S., Gu, J., Chen, J. (2005c). Gas hydrate fast nucleation from melting ice and quiescent growth along vertical heat transfer tube. *Science in China Series B-Chemistry*, 48(1), 75–82.

Xie, Y., Liu, D., Liu, N., Xiao, Y. (2008). Tetrahydrofuran hydrate formation-de-composition characteristics as cool storage medium. *Cryogenics and Refrigeration*, 918–921.

Xie, Y., Li, G., Liu, D., Liu, N., Qi, Y. (2009). Formation kinetics of isobutane hydrate as cool storage medium. In *8th IIR Conference on Phase Change Materials and Slurries for Refrigeration and Air Conditioning*, Karlsruhe, June 3–5.

Xie, Y., Li, G., Liu, D., Liu, N., Qi, Y., Liang, D., Guo, K., Fan, S. (2010). Experimental study on a small scale of gas hydrate cold storage apparatus. *Applied Energy*, 87(11), 3340–3346.

Xie, Y., Zhou, X., Xie, Z., Fang, Y. (2014). The CO_2 hydrates produced in a compression cycle used for cool storage on different initial pressure. In *11th IIR Gustav Lorentzen Conference on Natural Refrigerants: Natural Refrigerants and Environmental Protection*, Hangzhou, August 31–September 2.

Xu, H., Yang, R., Zhang, Y.P., Huang, Z., Lin, J., Wang, X. (2005). Thermal physical properties and key influence factors of phase change emulsion. *Chin. Sci. Bull.*, 50(1), 88–93.

Yamazaki, M., Sasaki, C., Kakiuchi, H., Osano, Y.T., Suga, H. (2002). Thermal and structural characterization of trimethylolethane trihydrate. *Thermochimica Acta*, 387(1), 39–45.

Yan, K.-L., Sun, C.-Y., Chen, J., Chen, L.-T., Shen, D.-J., Liu, B., Jia, M.-L., Niu, M., Lv, Y.-N., Li, N. et al. (2014). Flow characteristics and rheological properties of natural gas hydrate slurry in the presence of anti-agglomerant in a flow loop apparatus. *Chemical Engineering Science*, 106, 99–108.

Yang, S.O., Yang, I.M., Lee, C.S. (2000). Measurement and prediction of phase equilibria for water + CO_2 in hydrate forming conditions. *Fluid Phase Equilibria*, 175(1/2), 75–89.

Yang, D., Le, L.A., Martinez, R.J., Currier, R.P., Spencer, D.F., Deppe, G. (2008). Heat transfer during CO_2 hydrate formation in a continuous flow reactor. *Energy & Fuels*, 22(4), 2649–2659.

Yang, M., Song, Y., Jiang, L., Zhu, N., Liu, Y., Zhao, Y., Dou, B., Li, Q. (2013). CO_2 hydrate formation and dissociation in cooled porous media: A potential technology for CO_2 capture and storage. *Environmental Science & Technology*, 47(17), 9739–9746.

Yang, M., Song, Y., Jiang, L., Liu, Y., Wang, X. (2015). Behaviour of hydrate-based technology for H_2/CO_2 separation in glass beads. *Separation and Purification Technology*, 141, 170–178.

Youssef, Z., Delahaye, A., Huang, L., Trinquet, F., Fournaison, L., Pollerberg, C., Doetsch, C. (2013). State of the art on phase change material slurries. *Energy Conversion and Management*, 65(0), 120–132.

Youssef, Z., Hanu, L., Kappels, T., Delahaye, A., Fournaison, L., Zambrana, C., Pollerberg, C. (2014). Experimental study of single CO_2 and mixed CO_2 + TBAB hydrate formation and dissociation in oil-in-water emulsion. *International Journal of Refrigeration*, 46, 207–218.

Youssef, Z., Fournaison, L., Delahaye, A., Pons, M. (2019). Management of vapor release in secondary refrigeration processes based on hydrates involving CO_2 as guest molecule. *International Journal of Refrigeration*, 98, 202–210.

Zafara, S., Dincer, I., Gadalla, M. (2015). Experimental testing and analysis of R134a clathrates based PCMs for cooling applications. *International Journal of Heat and Mass Transfer*, 91, 756–765.

Zalba, B., Marin, J.M., Cabeza, L.F., Mehling, H. (2003). Review on thermal energy storage with phase change: Materials, heat transfer analysis and applications. *Applied Thermal Engineering*, 23, 251–283.

Zhang, P. and Ye, J. (2014). Experimental investigation of forced flow and heat transfer characteristics of phase change material slurries in mini-tubes. *International Journal of Heat and Mass Transfer*, 79, 1002–1013.

Zhang, P., Ma, Z., Wang, R. (2010a). Forced flow and convective heat transfer of phase change material slurry in the heat exchangers. In *International Conference on Energy Conversion and Application (ICECA 2001)*, Wuhan, June 17–20.

Zhang, P., Ma, Z., Wang, R. (2010b). An overview of phase change material slurries: MPCS and CHS. *Renewable and Sustainable Energy Reviews*, 14(2), 598–614.

Zhang, P., Shi, X.J., Ma, Z. (2013). Solid fraction determination in cold storage by tetra-n-butyl ammonium bromide clathrate hydrate slurry. *International Journal of Refrigeration*, 36(3), 809–819.

Zhang, Y., Li, X., Chen, Z., Xia, Z., Wang, Y., Li, G. (2017). Formation behavior and controlling factor of methane hydrate in porous media. *Energy Procedia*, 142, 4044–4049.

Zhou, H. and Ferreira, C.I. (2014). Heat transfer and flow characteristics during the formation of TBAB hydrate slurry. In *11th IIR Gustav Lorentzen Conference on Natural Refrigerants: Natural Refrigerants and Environmental Protection*, Hangzhou, August 31–September 2.

Zhou, H., de Sera, I.E.E., Ferreira, C.A.I. (2015). Modelling and experimental validation of a fluidized bed based CO_2 hydrate cold storage system. *Applied Energy*, 158, 433–445.

Zylyftari, G., Ahuja, A., Morris, J.E. (2014). Nucleation of cyclopentane hydrate by ice studied by morphology and rheology. *Chemical Engineering Science*, 116, 497–507.

8

Concentrated Solar Power Plants and Storage

Régis OLIVÈS
PROMES, CNRS, UPVD, Perpignan, France

8.1. Introduction

Concentrated solar power (CSP) plants use mirrors to concentrate sunlight on a receiver, which collects and transfers solar energy to a heat transfer fluid (HTF). This fluid can be used to provide heat to final applications or to generate electricity by means of classic steam turbines. Power plants may contain a heat storage system that frees generation from intermittences due to night or cloudy periods, ensures high-quality electricity (supply continuity, voltage quality, service quality) and increases the capacity factor of the power plant. These solar power plants require high direct solar irradiation. In terms of electricity supply, they are of strong interest for the countries in the "Sunbelt", located in the 40° North and South of the equator (Siegel 2012).

They are currently the second means for massive storage of electricity worldwide. While hydroelectricity provides most of this storage with about 180 GW, CSP plants with storage reached approximately 6 GW in 2021. The storage being fully integrated into the system, it makes a difference when compared to other renewable energy processes, such as photovoltaic or wind energy.

Heat and Cold Storage 1,
coordinated by Pierre ODRU, Elena PALOMO DEL BARRIO.

8.2. Concentrated solar power plants and storage

8.2.1. *General principles*

CSP plants concentrate direct solar radiation to increase the temperature of an HTF. This fluid supplies directly or indirectly a thermodynamic machine which, in turn, generates electricity via an alternator. In the global solar radiation comprising direct normal radiation and diffuse radiation, only direct radiation can be used by optical concentration devices.

There are four main technologies depending on their concentration mode:

– Linear concentration:

- parabolic troughs;

- linear Fresnel mirrors.

– Point concentration:

- heliostats associated with a tower;

- parabolic dish.

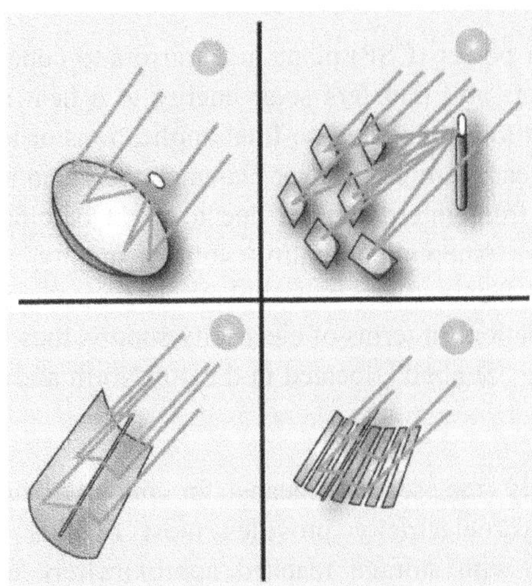

Figure 8.1. *Solar concentration principles. For a color version of this figure, see www.iste.co.uk/odru/heat1.zip*

The generally used thermodynamic cycles are Rankine, Brayton and Stirling cycles. Parabolic trough and linear Fresnel collector power plants, and also solar tower power plants, are often based on Rankine cycles and use steam. The Brayton cycle is mainly used in several cases of tower power plants where sufficiently high levels of temperature can be reached. It is implemented by means of a gas turbine that can be a hybrid solar/gas turbine. As for the parabolic concentrators, they are coupled with Stirling engines.

Considering the intermediate heat generation, these technologies can host a heat storage system.

A solar power plant has four main elements:

– the solar field that concentrates radiation;

– the solar receiver that converts this concentrated solar radiation into heat;

– the storage that accumulates heat and releases it later on;

– the power block that converts heat into electricity.

Figure 8.2 illustrates the coupling of these elements. It represents the configuration of the Thémis power plant located in the Eastern Pyrenees (France). With a 2.2 MW_e power, it operated between 1983 and 1986 and was a source of inspiration for many other power plants in the United States and Spain.

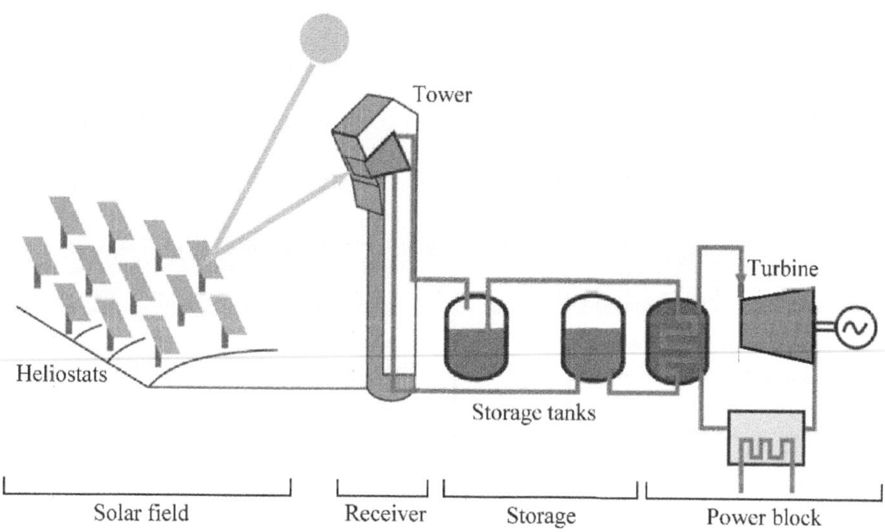

Figure 8.2. *Thémis solar tower power plant, Targasonne, Eastern Pyrenees. For a color version of this figure, see www.iste.co.uk/odru/heat1.zip*

Figure 8.3 presents the various types of solar power plants with their main characteristics. Among those, the concentration factor corresponds to the ratio between the power density obtained by the optical concentration system and the received power density of 1,000 $W \cdot m^{-2}$ taken as a reference. The type of concentration determines the levels of temperature, which in turn impose the thermodynamic cycle of the power block. Therefore, the storage implementation depends on the constraints induced by these two elements.

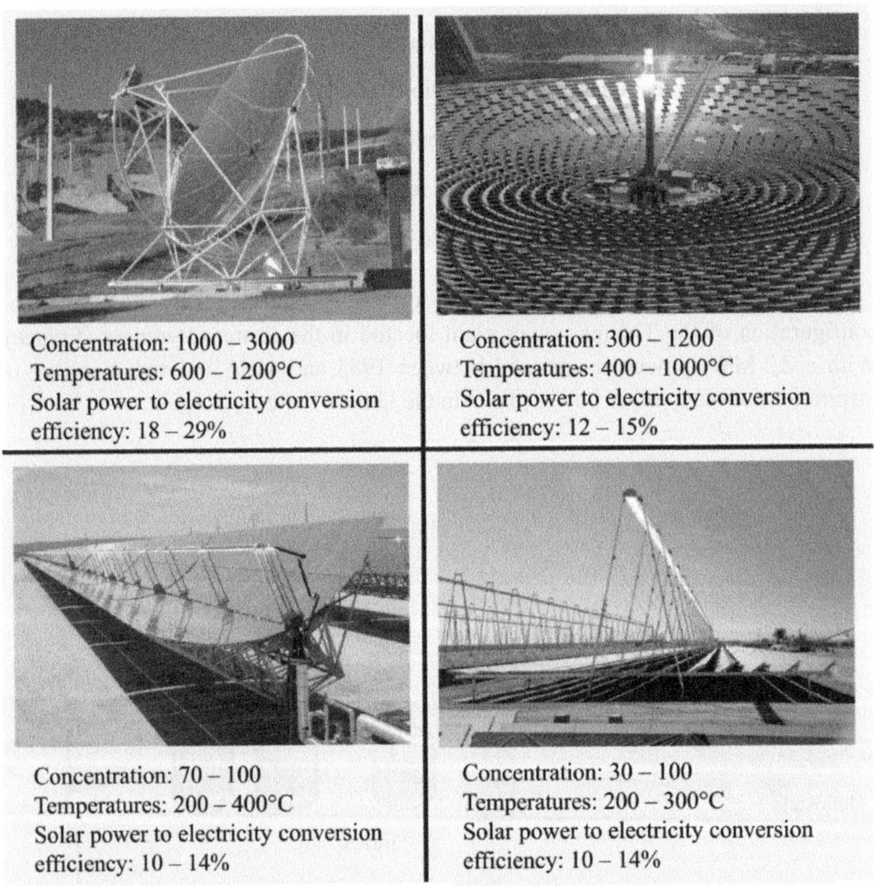

Figure 8.3. *Main types of concentrated solar power plants. For a color version of this figure, see www.iste.co.uk/odru/heat1.zip*

8.2.2. *Objectives and strategy*

Thermal energy storage systems in solar power plants contribute to the following functionalities:

– smoothing of fluctuations: storage acts as a buffer system to reduce the impact of the cloud cover, bad weather periods or day/night alternation. A low to moderate capacity storage may help in mitigating solar intermittencies by reducing the turbogenerator stop and restart phases, which have a negative effect on these performances;

– production time shift: the solar energy collected in excess is absorbed by the storage so that it can be released when demand reaches a peak. Hence, storage improves the capacity of the power plant to fit the demand of the electric grid (dispatchability);

– increase in the charge rate of the power plant (annual capacity factor): similarly to the above point, excess solar energy is collected to be able to generate electricity at further moments in time. By generating it continuously, the power plant can then ensure 24-h generation and thus considerably increase the annual number of hours of electricity generation.

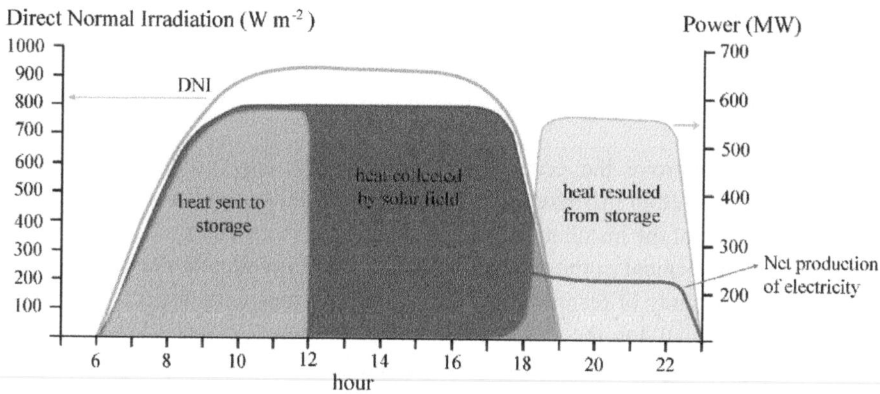

Figure 8.4. *Daily electric production of a solar power plant with thermal storage. For a color version of this figure, see www.iste.co.uk/odru/heat1.zip*

Storage becomes economically attractive if, by shifting production from low to high tariff periods, the increase in revenues covers the additional costs generated by storage. Obviously, a significantly higher charge rate may also reduce the cost of the electricity produced by the power plant. It is an additional argument for the investment in renewable energies and therefore a more environmentally friendly generation of electricity.

Figure 8.4 illustrates the case of a solar power plant that produces 250 MW electric power between 12 and 23 h. Hence, a part of the heat collected by the solar field is stored and released by the end of the day, when there is no insolation. Because of storage, a solar power plant has the capacity to significantly contribute to the massive storage of electricity.

However, for this purpose, the storage should meet a certain number of widely agreed upon criteria:

– high-energy density of the material used;

– efficient thermal exchange between the HTF and the storage;

– mechanical and chemical stability of the storage materials;

– chemical compatibility between the HTF, exchanger and storage materials;

– complete reversibility of the process for a large number of successive charge and discharge cycles;

– small thermal losses;

– low cost;

– minimized and controlled environmental impact.

In order to improve the economic value of the energy produced, operation strategies can be used to increase the efficiency of the subsystems and thus increase the electric power of the installation and the quantity of electricity produced. Storage appears as a key element in the production strategy and investment decision (Meyer et al. 2019). The study of technical, economic and environmental aspects of storage is therefore essential for the development of solar power plants (Achkari et al. 2020). Finally, storage in this type of power plant is a fundamentally differentiating element in relation to other renewable energy generation systems (Cabeza 2015).

8.2.3. *Global performances and efficiency*

The energy model of a solar power plant relies on the coupling between solar field, solar receiver, thermal storage and the power supply block (Figure 8.5). The input data of the model are the solar resource and, due to the concentration, direct normal irradiation should be known.

Input power is therefore linked to direct normal irradiation:

$$P_{sol} = DNI.S_c \quad [8.1]$$

where S_c is the surface of the solar field. This power is transmitted to the solar receiver according with the efficiency of the solar field η_c:

$$P_{cr} = \eta_c.P_{sol} = P_{sol} - P_{pc} \quad [8.2]$$

where P_{pc} is the power lost at the level of the solar field (optical losses). The receiver output power is divided between the power block (P_{rb}) and the thermal storage (P_{rs}):

$$\eta_r.P_{cr} = P_{cr} - P_{pr} = P_{rb} + P_{rs} \quad [8.3]$$

where η_r is the efficiency of the receiver and P_{pr} is the power lost at the level of the receiver. In order to generate electric power P_{el}, the power block requires power P_b issued directly from receiver P_{rb} and that from storage P_{sb}. Hence:

$$P_{sb} = \eta_s.P_{rs} \quad [8.4]$$

$$P_{el} = \eta_b.P_b = \eta_b.(P_{rb} + P_{sb}) \quad [8.5]$$

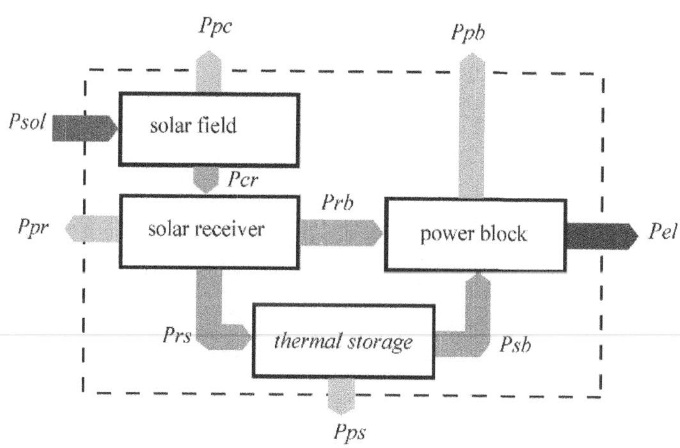

Figure 8.5. *Energy model of solar power plant. For a color version of this figure, see www.iste.co.uk/odru/heat1.zip*

The power blocks in power plants use standard thermodynamic cycles. Their efficiency depends in the first place on the temperature of the hot source:

$$\eta_b = \eta_{Carnot} . (1 - Dis) = \left(1 - \frac{T_o}{T_h}\right) . (1 - Dis) \qquad [8.6]$$

where η_{Carnot} is the Carnot efficiency, Dis stands for thermodynamic dissipations or irreversibility, T_o is the temperature of the environment and T_h the temperature of the hot source, reached by the HTF supplying the power block.

Hence, according to this simplified model, the amount of electricity produced by the installation is linked to the available annual direct normal insulation (Figure 8.6), annual efficiency of the solar field collection, thermal conversion efficiency of the solar receiver (radiation-heat), thermal losses, storage and thermodynamic cycle efficiencies (Winter et al. 1991).

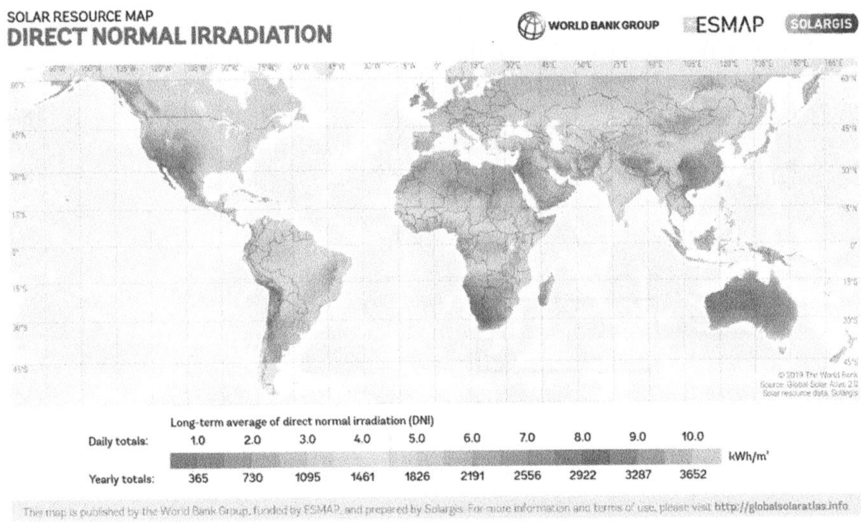

Figure 8.6. *Map of direct normal irradiation. For a color version of this figure, see www.iste.co.uk/odru/heat1.zip*

The efficiency of the storage unit depends on the type of storage and will be discussed in section 8.3. For direct production without storage, the efficiency of the installation is given by:

$$\eta = \eta_c . \eta_r . \eta_b \qquad [8.7]$$

while with storage, it is expressed as follows:

$$\eta = \eta_c . \eta_r . \eta_s . \eta_b \qquad [8.8]$$

The solar multiple (SM) is defined as the ratio between the thermal power recovered by the solar field and the thermal power entering the power block (Montes et al. 2009):

$$SM = \frac{P_c}{P_b} \qquad [8.9]$$

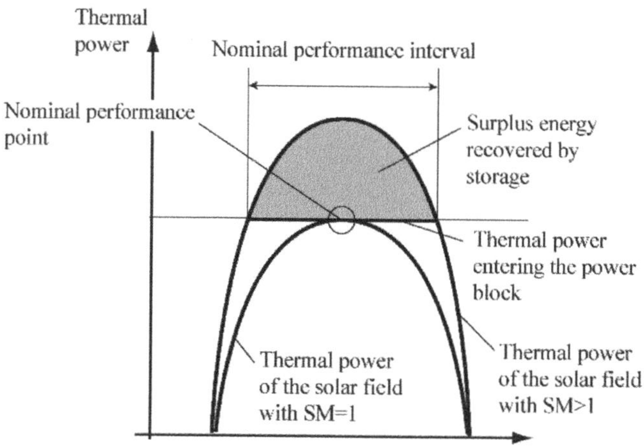

Figure 8.7. *Solar multiple*

In terms of nominal thermal power, this parameter represents the size of the solar field relative to the power of the supply block. An oversized solar power plant ($SM > 1$) produces more energy at full solar power than the amount the supply block can convert. Therefore, it requires the integration of a storage that absorbs this surplus energy.

The consequence of an increase in storage capacity would be an increase in the capacity factor of the power plant, that is, the quantity of energy produced during the year. The parameter SM and the capacity factor are therefore linked, as shown in Figure 8.8 (Winter et al. 1991). In other words, for a fixed nominal power installation, when the size of the solar field increases, the capacity factor increases and then saturates. A further increase in the annual production of electricity requires an increase in storage size.

The choice of the power plant technology, the type of fluid used and the mass flow rate of the fluid imposes the nominal input and output temperatures of the solar field and, consequently, the heat supplied by the solar field (Winter et al. 1991). The mass flow rate of the transfer fluid is adjusted throughout the day to ensure optimal operation within the expected temperature range. The temperature of the transfer fluid at the outlet of the solar field also determines the temperature at the storage inlet. Therefore, one of the main requirements for the selection of suitable storage is the nominal working temperature of the fluid in the solar field. Additionally, the maximal electric power, the operation mode desired for the storage system and the type of storage integration in the power plant should be considered.

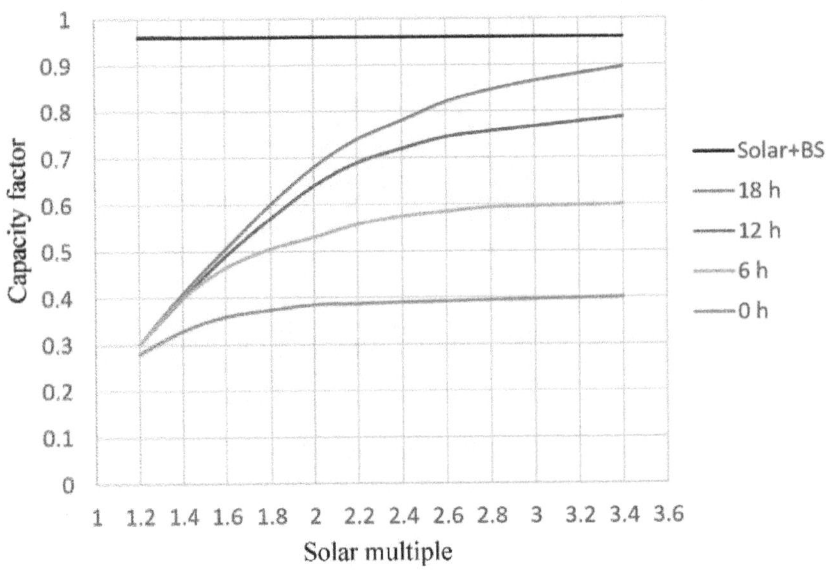

Figure 8.8. *Relation between the capacity factor and the solar multiple depending on the number of storage hours (solar+BS: solar with fossil fuel contribution) (according to Winter et al. (1991)). For a color version of this figure, see www.iste.co.uk/odru/heat1.zip*

Storage operation is therefore strongly dependent on the solar field and power block. It is important to note that the numerical simulation of the entire process requires the integration of models of each of the subsystems for the development of the strategy of operation, predicting production and improving the performances of the solar power plant (Kuravi et al. 2013).

8.3. Storage types

The performances of a storage system depend on components and on their respective performances. Besides the storage type, the design requires the selection of storage material, HTF and envelope material, while ensuring compatibility between the three. Moreover, the pumps for HTF circulation operating storage charge and discharge and, possibly, the heat exchangers (HEs) between the solar field and the storage, and between the storage and the power block, should be considered.

Storage systems are classified according to the storage mode – sensible heat, latent heat, thermochemical – and also according to their active or passive type.

In *active* storage, the storage medium serves also as HTF. If the storage uses a medium other than the HTF, it is known as *passive*.

The circulation is *direct* when the HTF is the same for the solar field and the storage and *indirect* if the fluid is different. In this latter case, heat is transferred through an HE.

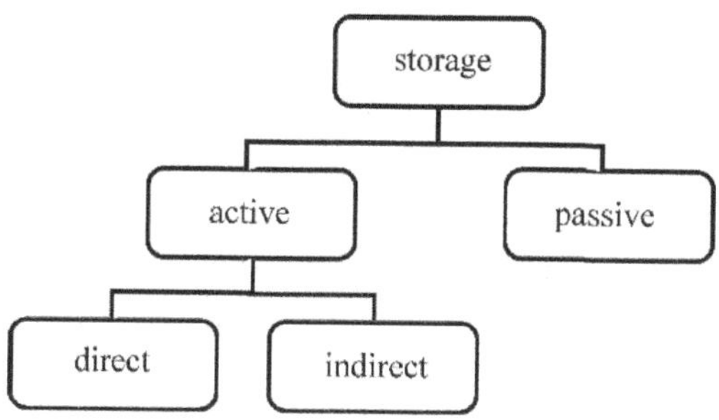

Figure 8.9. *Types of active or passive storage and direct or indirect circulation*

Each of these configurations has its own advantages and disadvantages compared to others. For example, the configuration of an active system with direct circulation allows cost containment while increasing the efficiency of the solar power plant by eliminating the HE between the solar field and the heat storage. The design of the set must then take into account the compatibility of the fluid with the set of components of the power plant. In the case of indirect circulation, the choice of each fluid is optimized depending on its application (heat transfer or storage). In terms of thermal performances, it is essential for the storage system to possess sufficiently high-energy density with an adequate charge and discharge rate. In the case of active storage systems, liquids are preferred to gases, as their energy density and thermal conductivity is higher than that of gases. However, the use of a liquid meeting all the conditions (temperature stable, chemically inert, noncorrosive, etc.) may be economically impossible. Similarly, the use of a solid, therefore passive material in a system, may lead to efficiency containment in terms of energy charge and discharge. To address these difficulties, conducting materials can be integrated in order to increase heat transfers and improve charge and discharge power.

8.3.1. *Sensible heat storage*

Sensible heat storage involves bringing a medium to a higher temperature when excess heat is available and cooling it when heat needs to be released. The amount of heat is associated with a variation in enthalpy:

$$Q = \Delta H = \int_{T_{ini}}^{T_{fin}} m.C_p.\mathrm{d}T = \int_{T_{ini}}^{T_{fin}} \rho.V.C_p.\mathrm{d}T \qquad [8.10]$$

where C_p is the specific heat capacity at constant pressure, when thermal expansion is ignored. It is important to note that for a gas, the enthalpy related to the variation of temperature at constant volume is given by:

$$Q = \Delta H = \int_{T_{ini}}^{T_{fin}} \rho.V.C_v.\mathrm{d}T \qquad [8.11]$$

Therefore, the amount of heat stored per unit volume is given by:

$$Q_V = \int_{T_{ini}}^{T_{fin}} \rho.C_p.\mathrm{d}T \qquad [8.12]$$

All the storage systems installed in large-scale CSP plants store energy using sensible heat. The most commonly used systems are active with indirect circulation based on molten salts flowing between two tanks (Figure 8.10). The storage fluid is therefore the same as the one flowing between the two tanks and exchanges heat with the HTF that flows in the solar field.

The Andasol power plant is a typical example of using a two-tank indirect sensible heat storage. This power plant with parabolic trough concentrators and electric power of about 50 MW has two tanks containing 28,500 tonnes of a binary mixture of molten salts (60% $NaNO_3$/40% KNO_3). The storage has a capacity of 1,010 MWh corresponding to 7.5 h of production at nominal power.

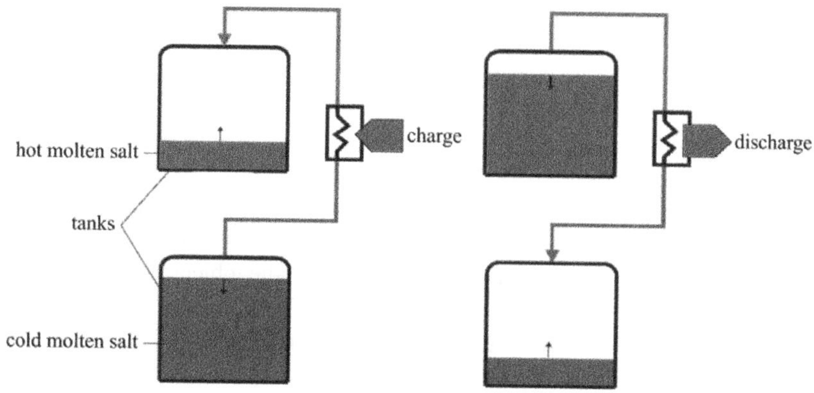

Figure 8.10. *Operation of a two-tank active storage system*

Figure 8.11. *Andasol power plant, Andalusia, Spain. For a color version of this figure, see www.iste.co.uk/odru/heat1.zip*

There are several examples of active storage with direct circulation. It is the case when the steam generated in the solar receiver is directly stored in a tank. This tank, known as an accumulator, contains liquid water reheated by this steam under pressure. If the system is charged with superheated steam, the temperature and pressure of the water in the tank increase, thus modifying the saturation state of the initial mass. The discharge occurs by reducing the pressure in the storage tank. As a result, saturated steam is generated, which decreases pressure during discharge. This technology used for short duration discharges (typically 30 min) is intended to smooth the insulation fluctuations. The storage in PS10 power plant in Seville relies on this principle (Figure 8.12).

Though there was a strong development of two-tank sensible heat storage systems with molten salt for solar power plants due to the simplicity of implementation, there is ongoing research to reduce the costs and environmental impact essentially related to molten salts. In order to reduce these costs, thermocline technology has been implemented and is currently the object of many research works. This technology, which is used in certain glass industries, for example, and more widely known as regenerator (Willmott 2001), consists of storing heat in a single tank. Hence, a HTF heated by the solar field is introduced at the top of the tank and charges the system. A thermal gradient forms between hot and cold zones, and it is known as thermocline. This zone separates the hot part from the cold part and avoids the homogenization of temperature throughout the tank; the amount of heat stored at high temperature is this way preserved. The discharge phase consists of injecting cold fluid by the bottom and recovering the heat of the fluid going out at the top of the tank. In order to preserve stratification, it is important to homogeneously inject the fluid all over the section.

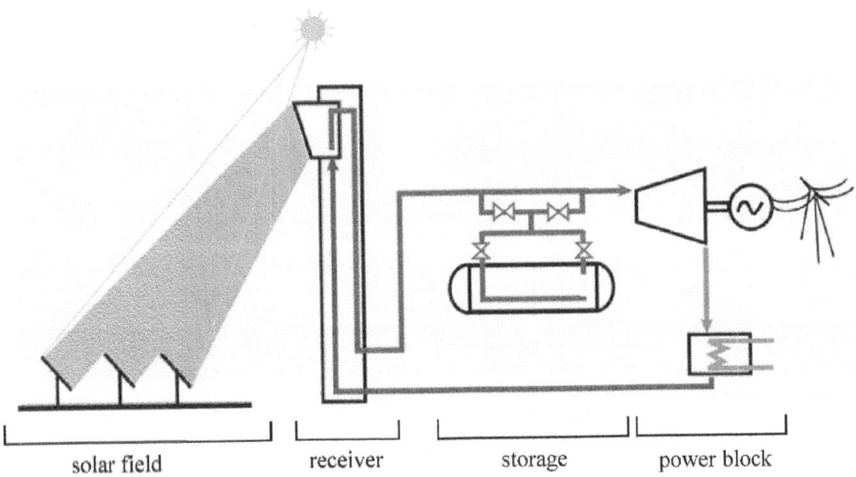

Figure 8.12. *Principle of PS10 power plant, Seville, Spain. For a color version of this figure, see www.iste.co.uk/odru/heat1.zip*

In order to reduce the quantity of fluid and therefore the cost of storage, the tank is filled with a solid material. The latter can be structured or unstructured. This configuration corresponds to a passive storage. It also involves two phases and relies on the existence of the thermocline. During the charge, the HTF brings the heat and transfers it to the solid that accumulates it. During the discharge, the HTF recovers the heat released by the solid. The general operation principle of the thermocline tank is therefore the following:

– A discharge consists of sending cold fluid to the bottom of the tank and extracting hot fluid from the top. The discharge is completed when the thermocline zone is extracted from the tank.

– A charge consists of injecting hot fluid into the top of the tank and extracting cold fluid from the bottom. The charge is completed when the thermocline zone is extracted from the tank.

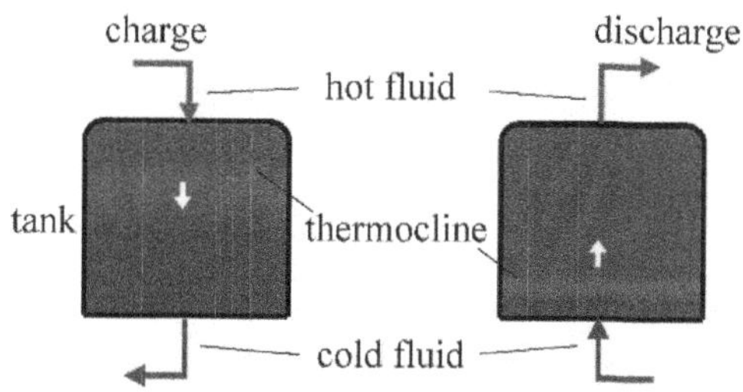

Figure 8.13. *Thermocline storage operation. For a color version of this figure, see www.iste.co.uk/odru/heat1.zip*

The first large-scale example was built at Solar One at Barstow (California) and operated between 1982 and 1986 (Friefeld et al. 1982; Faas et al. 1986). Initiated in 1979 by the DOE, Southern California Edison and the Los Angeles Department of Water and Power, the project consisted of the integration of a 170 MWh thermocline storage in the tower power plant with a nominal production of 10 MW_e. This power plant has been refurbished to allow a two-tank molten salt storage: Solar Two.

8.3.2. *Latent heat storage*

These systems use phase change materials (PCMs) during the exchange with the HTF. It is a matter of benefiting from the enthalpy associated with the phase transition: latent heat. This phase change phenomenon may a priori be a solid/solid, liquid/vapor or solid/liquid transition. The solid/liquid transition is the most studied due to the small variation of volume at the passage from one phase to another compared to the liquid/vapor transition and the larger latent heat compared to the solid/solid transition. The change of state of a pure material is therefore characterized by the transformation, at constant temperature, from one phase to another. This transformation is controlled by the variation of free energy:

$$\Delta G = \Delta H - T_m.\Delta S \quad [8.13]$$

where T_m is the transition temperature and S is the entropy. At equilibrium, $\Delta G = 0$:

$$\Delta H = T_m.\Delta S \quad [8.14]$$

For a fixed transition temperature T_m, a large variation of entropy corresponds to a large latent heat. The amount of heat stored during phase change is given by:

$$Q = \Delta H = m.L_m \quad [8.15]$$

The stored energy is thus expressed in a general manner by the following relation:

$$Q = \int_{T_{ini}}^{T_m} \rho.V.C_p.\mathrm{d}T + \rho.V.L_m + \int_{T_m}^{T_{fin}} \rho.V.C_p.\mathrm{d}T \quad [8.16]$$

The advantage of latent heat storage is that it essentially occurs at constant temperature and it involves a larger amount of energy per unit mass compared to sensible heat storage. This characteristic may prove essential in solar fields constrained by limit temperatures related to the transfer fluid and to the fluid circulating in thermodynamic cycles. In the case of steam generation for the turbine, the integration of a latent heat storage is relevant with respect to the transfer conditions in the HE. As the storage capacity of a latent heat storage system is controlled by the phase change enthalpy in addition to the specific heat of the material, the main interest is the potential reduction of the thermal storage volume.

One of the prototypes tested in the European project DISTOR (2004/2007) was a latent heat storage module based on a sandwich configuration and with 100 kW_{th} nominal power (Medrano et al. 2010). This prototype was installed at the Plataforma Solar de Almería (PSA) to be tested under actual operation conditions with steam generated by the DISS loop (parabolic trough concentrators).

This prototype consisted of a bundle of 36 parallel pipes comprising six tubes arranged in six sets. Figure 8.14 presents a schematic representation of the prototype. The bundle of pipes is placed in a block of PCM, whose conductivity is increased by graphite fins. A diagram of the sandwich PCM/fins configuration is shown in Figure 8.15. PCM is an eutectic mixture of $54\%_m$ $KNO_3/46\%_m$ $NaNO_3$, and the conducting fins are expanded graphite foils of 490 × 490 mm^2 and 1 mm thickness, at 10 mm distance from each other and perpendicular to the tubes. The storage prototype was filled with 2,100 kg PCM with a melting point at 221°C and a latent heat of about 100 kJ·kg^{-1}. This means that the maximal latent energy stored or released is about 210 × 10^3 kJ or 58 kWh_{th} and the expected charge/discharge time is about 35 min for a nominal power of 100 kW.

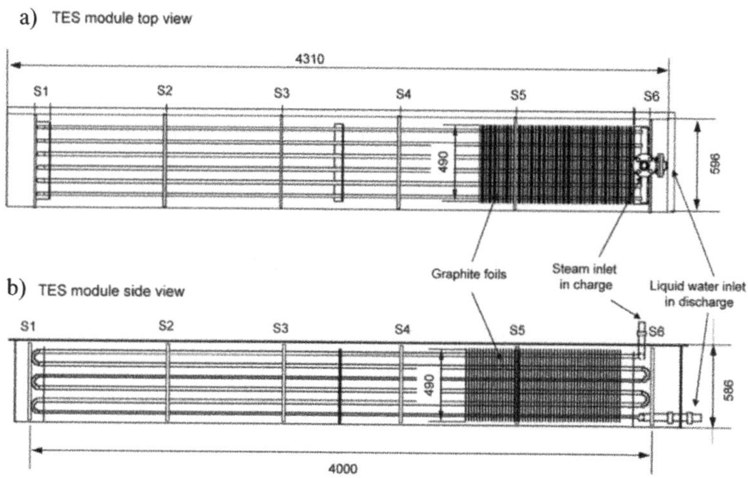

Figure 8.14. *Schematic representation of the latent heat storage prototype – DISTOR Project: (a) top view; (b) side view (dimensions in mm) (source: Medrano et al. 2010)*

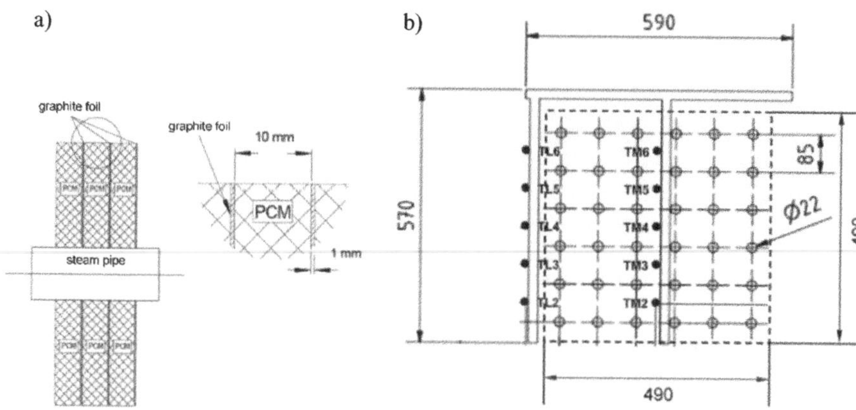

Figure 8.15. *Schematic representation of the prototype: (a) sandwich PCM/ graphite foil configuration; (b) layout of the bundle of pipes inside the prototype (dimensions in mm) (source: Medrano et al. 2010)*

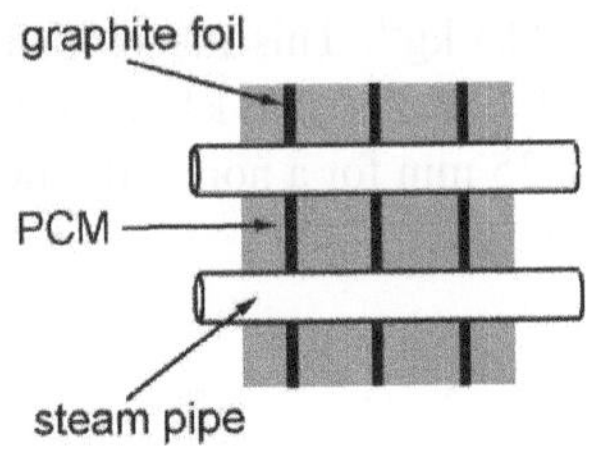

Figure 8.16. *Composite configuration with graphite fin (source: Medrano et al. 2010)*

Other configurations integrating PCMs have been designed and tested. It is worth considering the example of storage combining latent heat and sensible heat as shown in Figure 8.17. The interest in this type of configuration lies in the stabilization of the outlet fluid temperature during charge and discharge.

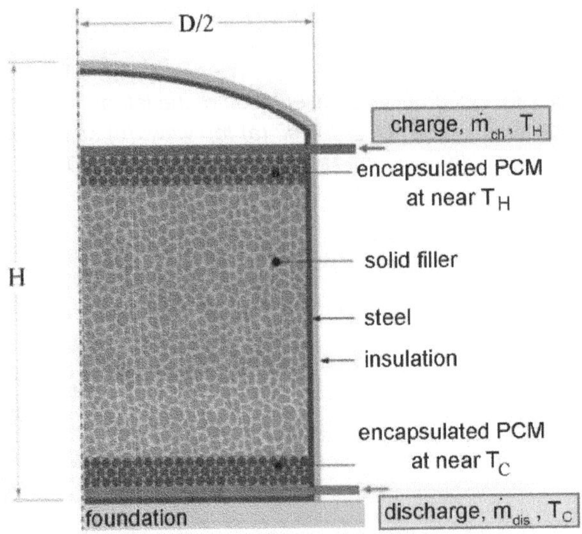

Figure 8.17. *Three layer latent/sensible/latent configuration (source: Galione et al. 2015). For a color version of this figure, see www.iste.co.uk/odru/heat1.zip*

8.3.3. *Thermochemical storage*

Despite its potential capacity to store more energy than the sensible or latent heat systems, this storage technology is still underdeveloped. Most studies have been conducted at the laboratory scale. These systems rely on the heat released or absorbed during reversible chemical reactions (Pardo et al. 2014):

$$A + \Delta H_r \Leftrightarrow B + C \qquad [8.17]$$

Therefore, during the charge phase, the heat brought to material A leads to the dissociation into products B and C. The reaction is therefore endothermic. During the discharge phase, the reverse reaction occurs: products B and C associate to form the initial product A. This reaction then releases heat; it is exothermic. The stored heat is thus directly linked to the enthalpy and to the rate of the chemical reaction:

$$Q = a_r . m_r . \Delta H_r \qquad [8.18]$$

where a_r is the fraction having reacted, m_r is the mass of the reactive material and ΔH_r is the heat of reaction per unit mass.

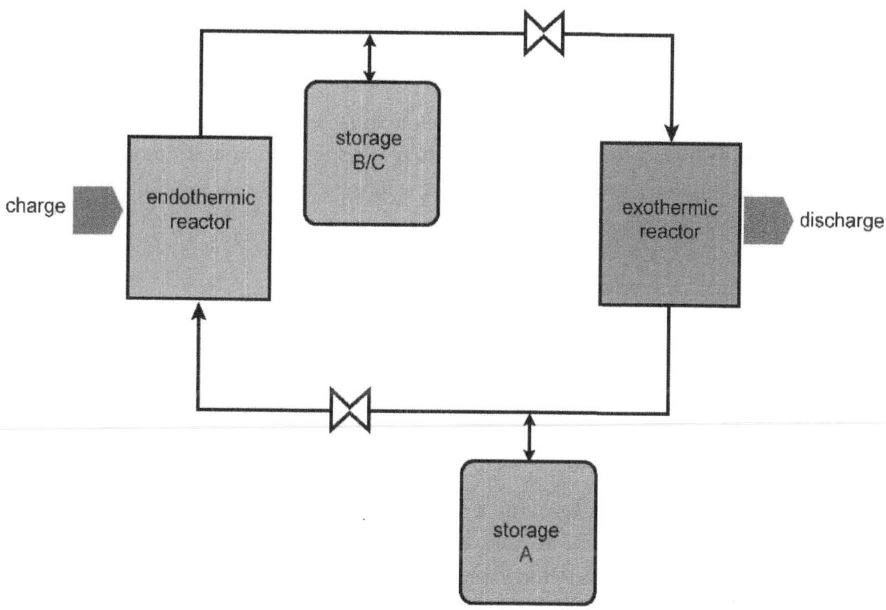

Figure 8.18. *Principle of thermochemical storage. For a color version of this figure, see www.iste.co.uk/odru/heat1.zip*

The following main reactions are listed:

$$MgH_2 + \Delta H_r \Leftrightarrow Mg + H_2 \quad [8.19]$$

$$PbCo_3 + \Delta H_r \Leftrightarrow PbO + CO_2 \quad [8.20]$$

$$CaCo_3 + \Delta H_r \Leftrightarrow CaO + CO_2 \quad [8.21]$$

$$Mg(OH)_2 + \Delta H_r \Leftrightarrow MgO + H_2O \quad [8.22]$$

$$2NH_{3(g)} + \Delta H_r \Leftrightarrow N_{2(g)} + 2H_{2(g)} \quad [8.23]$$

$$NH_4HSO_4 + \Delta H_r \Leftrightarrow NH_3 + H_2O + SO_3 \quad [8.24]$$

$$2Co_3O_4 + \Delta H_r \Leftrightarrow 6CoO + O_2 \quad [8.25]$$

$$2BaO_2 + \Delta H_r \Leftrightarrow 2BaO + O_2 \quad [8.26]$$

$$CH_4 + CO_2 + \Delta H_r \Leftrightarrow 2H_2 + 2CO \quad [8.27]$$

$$C_6H_{12} + \Delta H_r \Leftrightarrow C_6H_6 + 3H_2 \quad [8.28]$$

$$2SO_3 + \Delta H_r \Leftrightarrow 2SO_2 + O_2 \quad [8.29]$$

The main advantages of these thermochemical systems are high-energy densities of the order of 500 $kWh \cdot m^{-3}$, the quasi-constant charge or discharge temperature and the fact that in time these materials can store without thermal loss and therefore they can be used for long time storage.

The first stage of development of a thermochemical system is therefore the selection of the reaction and the study of its chemical characteristics, such as reversibility, reaction rate, operation conditions (P and T) and kinetic properties.

Pardo et al. (2014) present the advantages and disadvantages of each of these reactions. The data in the literature are listed in Table 8.1.

High enthalpies of reaction and a low molar volume product are required to maximize storage capacity. The two reactions must be fully reversible, without secondary reactions, and must have high efficiencies in order to use the materials over a long period or on a significant number of cycles. The two reactions must have sufficiently rapid kinetics so that the absorption of solar energy and the release of heat can be done over compatible durations. The chemical compounds of the two reactions must be easy to handle. Once stored, the chemical compounds should not react with their environment. The analysis of reactions proves to be necessary in order to control the thermochemical process. Finally, costs must be controlled and remain as low as possible.

Though there is a variety of thermochemical processes, no system has yet been implemented at the industrial scale. Nevertheless, several demonstrators have been tested. The most studied thermochemical system is based on the reaction with ammonia. In the presence of solar energy, ammonia (NH_3) dissociates into H_2 gas and N_2 gas that can be stored for an unlimited period in a pressurized tank.

During heat production, gases react to resynthesize NH_3 according to an exothermic process similar to the Haber–Bosch industrial process. The standard power of the tested systems is about 10 kW_{th}.

A pilot thermochemical storage by oxidation-reduction was implemented and used, in a representative manner, in a solar tower power plant (Tescari et al. 2017). The storage unit consists of inert honeycomb substrates (cordierite) coated with 88 kg of redox active material (cobalt oxide). The average energy that can be stored or released in a temperature range of about 700–1,000°C corresponds to 47 kWh for the full system.

Schmidt et al. (2014) conducted conclusive tests on the calcium hydroxide system [$Ca(OH)_2$], proving dehydration and rehydration (450°C and 550°C, respectively) of 20 kg of $Ca(OH)_2$. They reached a maximal thermal power of 7.5 kW_{th} with an air output temperature of 450°C during 35 min. For an operating pressure of 450 kPa, the output temperature may reach 600°C. Based on these results, they proposed and thermodynamically analyzed a concept of integration in a solar power plant for a storage up to 15 h.

Therefore, thermochemical storage systems have several advantages. Their energy densities are 5–10 times higher than those of latent or sensible heat storage systems, respectively. The storage duration is theoretically unlimited; as there is no thermal loss during storage, the products are stored at ambient temperature.

Hence, these systems are promising for the storage of thermal solar energy over a long period of time. Nevertheless, in contrast to the sensible and latent heat storage systems, there is only little feedback on large-scale thermochemical storage.

Reaction	Phase	Enthalpy of reaction	Temperature operating conditions	Energy density
Hydride system $MgH_2 + \Delta Hr \Leftrightarrow Mg+H_2$	Solid-gas	ΔHr=75 kJ/mol	Charge at 653 K Discharge at 503 K	580 kWh/m^3 (MgH_2) 0.80 kWh/kg (MgH_2)
Carbonate system $PbCO_3 + \Delta Hr \Leftrightarrow PbO+CO_2$	Solid-gas	ΔHr=88 kJ/mol	Charge at 723 K Discharge at 573 K	303 kWh/m^3 ($PbCO_3$) 0.80 kWh/kg ($PbCO_3$)
$CaCO_3 + \Delta Hr \Leftrightarrow CaO+CO_2$	Solid-gas	ΔHr=178 kJ/mol	Charge at 1,133 K Discharge at 1,153 K	692 kWh/m^3 ($CaCO_3$) 0.49 kWh/kg ($CaCO_3$)
Hydroxide system $Mg(OH)_2 + \Delta Hr \Leftrightarrow MgO+H_2O$	Solid-gas	ΔHr=81 kJ/mol	Charge at 423 K Discharge at 373 K	388 kWh/m^3 ($Mg(OH)_2$) 0.39 kWh/kg ($Mg(OH)_2$)
$Ca(OH)_2 + \Delta Hr \Leftrightarrow CaO+H_2O$	Solid-gas	ΔHr=104 kJ/mol	Charge at 723 K Discharge at 298–673 K	437 kWh/m^3 ($Ca(OH)_2$) 0.39 kWh/kg ($Ca(OH)_2$)
Ammonium system $NH_4HSO_4 + \Delta Hr \Leftrightarrow NH_3+H_2O+SO_3$	Liquid-gas	ΔHr=336 kJ/mol	Charge at 1,200 K Discharge at 700 K	860 kWh/m^3 (NH_4HSO_4) 0.81 kWh/kg (NH_4HSO_4)
$2NH_{3(g)} + \Delta Hr \Leftrightarrow H_{2(g)}+2H_{2(g)}$	Gas	ΔHr=66.9 kJ/mol (charge) ΔHr=53 kJ/mol (discharge)	Charge at 723 K P=9–150 bar Discharge at 723 K P=10–300 bar	745 kWh/m^3 ($NH_{3(l)}$) 1.09 kWh/kg ($NH_{3(l)}$)

Table 8.1. *Data on the reactions studied for thermochemical storage at temperatures ranging between 573 and 1,273 K (source: Pardo et al. 2014)*

8.3.4. *Materials and fluids*

8.3.4.1. *Types of materials and properties*

The essential subcomponent of any thermal energy storage is the storage medium (Winter et al. 1991; Gil et al. 2010; Khare et al. 2013; Cabeza 2015). It must be carefully chosen so that its thermophysical properties, reactivity and thermal and chemical stability contribute to the robustness over several thousand thermal cycles. Besides the cost, these criteria play an essential role in the selection and significantly contain the list of media that may be used for the storage. The main thermophysical properties to be considered in the selection are the density, specific heat capacity, thermal conductivity, thermal expansion coefficient and the stability to cycling. Additionally, availability, production methods and the costs of these materials should be considered.

It is important to note the diversity of fluids and materials used for storage in solar power plants: steam, molten salts (nitrates), synthetic oils, vegetable oils, natural rocks or mixture of rock and gravel, sand, concrete, ceramics resulting from waste recycling, etc. Table 8.2 presents the main solid and liquid materials used for the sensible heat storage (Khare et al. 2013).

The fluids used in solar power plants are synthetic oils, molten salts, air and steam. Synthetic oils such as *Therminol VP-1* can generally be used up to 400°C and have low viscosity. They have fewer solidification problems (Tmin = 12°C) and can therefore be cooled down to the ambient temperature, if necessary. They are however relatively expensive, flammable and must be used under pressure (10 bars at 400°C). If they are used as storage fluid, a huge mass of fuel product must be used at the center of a tank. In terms of safety, this does not seem necessarily appropriate. Research works have been conducted on vegetable oils aiming at replacing synthetic oils used in lower temperature ranges with a lower environmental impact (Hoffmann et al. 2018, 2019).

Molten salts are used up to 600°C and have a volumetric energy capacity higher than that of oils. They are also less expensive than the latter and are not flammable. In the case of nitrate salts used in large quantity with oil at high temperature, regulations are however strongly binding. Moreover, due to a relatively high viscosity, the electric consumption of the circulation system may prove to be difficult to ignore. Furthermore, their solidification point is significantly above the ambient temperature (between 130 and 220°C depending on salts) and therefore requires a system adapted to avoiding their solidification (electrical tracing, for example). Finally, since they are corrosive, there is a higher risk of pipes getting deteriorated. This problem is all the more critical as the length of pipes may reach several kilometers in the parabolic trough solar fields. In most power plants, these molten salts can be used both as HTF and as storage fluid (Cabeza 2015).

Storage material	Temperature		Density (kg/m^3)	Thermal conductivity (W/m K)	Thermal capacity (kJ/kg K)
	Low (°C)	High (°C)			
Sand-mineral oil	200	300	1,700	1	1.3
Reinforced concrete	200	400	2,200	1.5	0.85
NaCl (solid)	200	500	2,160	7	0.85
Cast iron	200	400	7,200	37	0.56
Silica refractory brick	200	700	1,820	1.5	1
Magnesia refractory brick	200	1,200	3,000	1	1.15
Hitec solar salt	120	133	1,990	0.6	
Mineral oil	200	300	770	0.12	2.6
Synthetic oil	250	350	900	0.11	2.3
Silicon oil	300	400	900	0.1	2.1
Nitrite salt	250	450	1,825	0.57	1.5
Nitrate salt	265	565	1,870	0.52	1.6
Carbonate salt	450	850	2,100	2	1.8
Sodium liquid	270	53	850	71	1.3
Silicon carbide	200	1,400	3,210	3.6	1.06
SiO_2 (cristobalite)	200	1,200	2,350	0.92	1.13

Table 8.2. *Solid and liquid materials for sensible heat storage (source: Khare et al. 2013)*

The air has the advantage of being readily available at a minimal cost, its maximal working temperature being above 1,000°C. Therefore, it will be preferred in systems reaching high temperatures, such as the tower power plants. However, since it does not have a high volumetric storage capacity, it must be associated with a material having this property. It is also necessary to make it circulate at high flow rates, which adds to the costs of the circulation and compression systems.

Liquid water is one of the best storage media, with very high volumetric capacity and low cost. However, it cannot exceed 100°C atmospheric pressure, the threshold beyond which it transforms into steam. While it can be considered up to about 120°C and under pressure, water cannot meet the conditions imposed by thermodynamic cycles adapted to the heat into electricity conversion. In the case of liquid/vapor transition, energy is latently stored, a process that cannot take place beyond 374.15°C and 221.2 bars (critical point). Due to its very low density, it is difficult to use steam to store sensible heat. Steam accumulation-based storage can only take place at very high pressure and with limited storage durations.

Even though molten metals such as sodium have interesting thermophysical properties, they are not used in solar power plants because of their very strong reactivity, risks of explosion and especially their costs.

In the case of thermocline storage, the solid materials employed to store heat must allow cost reduction. Considering the amounts of energy involved, this storage requires a large amount of materials. Considering a massive deployment of this type of technology, it becomes important to pay attention to the availability of resources and the impacts generated by their use. Moreover, according to the IEA scenario, CSP plants are expected to provide approximately 4.5% of the electric energy production by 2050. Let us consider as reference the Andasol power plant of 50 MW_e and the equivalent at nominal production of 7.5 storage hours ensured by 28,500 tonnes of molten salts. By 2050, the equivalent of 20–30 power plants should be built annually and an amount of 570,000–860,000 tonnes of salt should be produced each year. But the annual production of nitrate salt is currently 800,000 tonnes and most of it is used for fertilizers (Py et al. 2013). The limits of such a resource are therefore obvious.

Natural rocks are potential candidates. They have thermophysical properties that are well suited for the storage with air as HTF (Hänchen et al. 2011). More precisely, mining sub-products may be a major source of profitable and ecological storage materials (Nahhas et al. 2020).

Industrial waste and co-products prove to be interesting materials for the preservation of virgin raw materials resources (Py et al. 2011; Faik et al. 2012; Calvet et al. 2013; Gutierrez et al. 2016; Lalau et al. 2016). As secondary raw material, they are a relatively abundant and cheap resource that offers a new cycle to materials.

Industrial waste and co-products can be used in large quantity, at low cost and reduced environmental impact. Therefore, they are relevant candidates for large-scale use. It is a matter of positioning them among the set of materials and defining them in terms of thermophysical properties.

Table 8.3 presents a quite large range of sub-products and materials resulting from waste.

Recent studies consider the use of airborne particles in order to have the advantages of this gas (low cost, no temperature limits) in addition to significant energy density (high volumetric particle storage capacity). This is a very promising perspective, but it is at an early stage (Zhang et al. 2017).

Latent heat storage uses families of inorganic salts. Table 8.4 presents the set of materials considered.

In the case of thermochemical storage, the materials and reactions tested or considered are shown in Table 8.5.

The thermophysical properties presented in the previous tables prove to be important for the selection of materials and the design of the system. The energy density determines quite obviously the storage capacity, while the thermal conductivity determines the charge and discharge power. Indeed, the conductivity controls the heat transfer dynamics and the system performances.

In the case of sensible heat storage on solid materials, the dynamic aspect of storage is then governed by the power with which heat is accumulated or released. This power depends on the quality of the exchange between the HTF and the storage material and on the transfer within the material. They must also involve an exchange surface facilitating the transfers between the material and the HTF. This depends then on the geometry and the presence of an encapsulation or systems that increase transfers, such as the fins.

Industrial materials (co-products or waste)	T(°C)	Density ($kg\ m^{-3}$)	Thermal capacity ($kJ\ kg^{-1}\ K^{-1}$)	Thermal conductivity ($W\ m^{-1}\ K^{-1}$)
Inert asbestos waste	20 – 1,000	3120	0.8 – 1.034	2.1 – 1.4
Fly ashes	20 – 1,100	2,600 – 2,962	0.714 – 1.3	1.16 – 1.59
Slags	200 – 750	2,800 – 3,844	0.71 – 0.97	1 – 2.2
Cu refining co-products	100 – 500	2,700 – 3,600	0.57 – 1.23	1.4 – 1.6
NaCl	20 – 400	1,384	0.738	0.33
Demolition waste	20 – 750	2,730	1.28 – 1.46	0.53

Table 8.3. *Materials resulting from waste for sensible heat storage (source: Gutierrez et al. 2016)*

Materials	Phase change temperature	Latent heat	Density	Energy density	Thermal conductivity	Manufacturer
(% mass)	(°C)	(kJ/kg)	(kg/m^3)	(kJ/m^3)	(W/m/K)	–
$MgCl_2$-$6H_2O$)	117	168.6	1,450 (liquid, 120°C)	244,470 (liquid, 120°C)	0.570 (liquid, 120°C)	–
–	–	–	1,569 (solid, 20°C)	264,533 (solid, 20°C)	0.694 (solid, 90°C)	–
$NaNO_3$–KNO_3 (50:50)	220	100.7	1,920	193,344	0.56	–
KCl–$ZnCl_2$ (68.1:31.9)	235	198	2,480	491,040	0.8	–
LiCl–LiOH (37:67)	262	485	1,550	751,750	1.1	–
$ZnCl_2$	280	75	2,907	218,025	0.5	–
H300	302	130	1,900	–	–	pcmproducts PlusICE
$NaNO_3$	308	199	2,257	449,143	0.5	–
NaOH	318	165	2,100	346,500	0.92	–
KNO_3	336	116	2,110	244,760	0.5	–
NaCl–KCl (58:42)	360	119	2,084.4	248,044	0.48	–
KOH	380	149.7	2,044	305,987	0.5	–
H395	395	215	2,330	–	–	pcmproducts PlusICE
H430	430	125	2,160	–	–	pcmproducts PlusICE
$MgCl_2$–NaCl (38.5:61.5)	435	351	2,480	870,480	–	–
Na_2CO_3–Li_2CO_3 (56:44)	496	370	2,320	858,400	2.09	–

Materials	Phase change temperature	Latent heat	Density	Energy density	Thermal conductivity	Manufacturer
H500	500	300	2,220	–	–	pcmproducts PlusICE
H535	535	130	2,320	–	–	pcmproducts PlusICE
H610	610	410	2,070	–	–	pcmproducts PlusICE
NaF–MgF_2 (75:25)	650	860	2,820	2,425,200	1.15	
H650	652	300	2,450	–	–	pcmproducts PlusICE
H700	699	250	2,410	–	–	pcmproducts PlusICE
$MgCl_2$	714	452	2,140	967,280	–	
H755	755	466	2,160	–	–	pcmproducts PlusICE
LiF–CaF_2 (80.5:19.5)	767	816	2,390	1,950,240	1.70 (liquid)	–
–	–	–	–	–	3.8 (solid)	–
NaCl	800	492	2,160	1,062,720	5	–
Na_2CO_3	854	275.7	2,533	698,348	2	–
H885	885	236	2,290	–	–	pcmproducts PlusICE
K_2CO_3	897	235.8	2,290	539,982	2	

Table 8.4. *Phase change materials for latent heat storage*

Compound	Reaction	Energy density	Reaction temperature (°C)
Ammonia	$NH_3 \leftrightarrow 1/2N_2 + 3/2H_2$	67 kJ mol^{-1}	400 – 500
Methane / water	$CH_4 + H_2O \leftrightarrow CO + 3H_2$	not available	500 – 1,000
Calcium hydroxide	$Ca(OH)_4 \leftrightarrow CaO + H_2O$	3 GJ m^{-3}	500
Calcium carbonate	$CaCO_3 \leftrightarrow CaO + CO_2$	4.4 GJ m^{-3}	800 – 900
Iron carbonate	$FeCO_3 \leftrightarrow FeO + CO_2$	2.6 GJ m^{-3}	180
Metal hydride	$MxH_2 \leftrightarrow MyH_2 + (x-y)H_2$	4 GJ m^{-3}	200 – 300
Magnesium hydroxide	$MgO + H_2O \leftrightarrow Mg(OH)_2$	3.3 GJ m^{-3}	250 – 400
Methanolation/demethanolation	$CH_3OH \leftrightarrow CO + 2H_2$	not available	200 – 250

Table 8.5. *Thermochemical materials and reactions (source: Pardo et al. 2014)*

The design of storage systems requires good knowledge of the properties of the materials and fluids used. Their properties have been characterized by many studies. Nevertheless, a difficulty may appear when the use of the materials tested in the laboratory is upgraded to industrial scale. Differences related to purity, the size of measured samples, manufacturing techniques, etc., can be noted. It is therefore essential to characterize the materials effectively used in solar power plant storage. Several cautionary points can be mentioned. For example, in the case of natural rocks, the variability of themophysical properties should be verified according to the origin and their resistance in time. For binary mixtures, segregation should be eliminated. PCMs must be validated on a scale as close as possible to the industrial one. The ageing of organic fluids during thermal cycles may lead to modifications of flow-related properties.

Similar to other systems, besides the sometimes prohibitive costs when used in large quantity, storage materials may have limitations to heat transfer. They can also be subjected to mass transfer limitations and to chemical stability or reversibility problems during cycling. Thermally performing materials can be obtained by using various transfer intensification techniques. This may consist of improving the thermal conductivity of the material or increasing the exchange surface between the material and the HTF. A possibility taken into consideration is to elaborate composite materials associating the strong thermal capacity of a first component with the strong conductivity of a second one in order to bring them closer to an effusivity of the order of or above 5,000 $J \cdot s^{-1/2} \cdot m^{-2} \cdot K^{-1}$.

In the case of latent heat storage, Py et al. (2001) and Mehling and Cabeza (2008) show that by increasing the thermal conductivity of the PCM, the heat transfer is intensified, allowing a larger distance between the heat transfer tubes and therefore a reduction of the necessary number of tubes. For example, for a 950 MWh_{th} system, a slight increase in thermal conductivity from 1 to 1.8 $W \cdot m^{-1} \cdot K^{-1}$ reduces the total required length of the tube by 46% (Laing et al. 2006). An analytical technique can be used to determine the required thickness of the PCM around a heat transfer tube for two PCMs of different latent heat (in the studied case, 100 $kJ \cdot kg^{-1}$ and 300 $kJ \cdot kg^{-1}$). For the same diameter of the external tube and the same thickness of the wall, the PCM with the higher latent heat requires a thickness of 9 mm, while the other PCM requires a thickness of 20 mm to store the same amount of energy. An improvement of the solidification times was noted when thermal conductivity increases to 5 $W \cdot m^{-1} \cdot K^{-1}$. Above this value, there is no significant improvement (Mehling and Cabeza 2008). Thermal conductivity can be significantly improved using graphite, metal fibers or foams.

Heat transfer can also be improved by using heat pipes between the storage material and the transfer fluid. Heat pipes have high efficiencies and can be designed in a wide range of shapes and for passive operation over a range of specific temperature.

Another technique to improve the heat transfer capacity of PCM involves the use of macro- or micro-encapsulation. Besides transfer intensification, encapsulation reduces the reactivity of PCM with the external environment and controls the variations in volume of the storage material during the phase change. Macro-encapsulated PCMs are then used as packed beds. Micro-encapsulation can also be considered in a host matrix. The applicability of these methods at high temperature is still a subject of research.

8.3.4.2. *Stability and compatibility*

As already mentioned, reactivity, thermal stability and chemical stability are important criteria for the selection of materials and fluids. The rigorous evaluation of these criteria may take time with respect to cyclical nature and long lifetime of storage systems. Ageing studies depend on testing conditions and on the methods used to extrapolate the results obtained on necessarily reduced durations.

HTFs must be compatible with the containment chamber and storage materials in the working temperature range. Thermal oils are preferred for use as HTF, as the constraints related to molten salts are too significant for a power plant with pipes of several kilometers (corrosion, operating cost of the circulation system given the high viscosity of salts). Molten salts are preferred for storage applications as they allow a decrease in the size of the tank and of the costs for an equivalent stored energy (higher volumetric capacity), but also because (unlike oils) they do not pose fire risks.

The commonly used organic oils are Dowtherm A, Therminol VP1 or Jarrytherm oils. They tend to decompose at high temperature (around 380-400°C). Inorganic salts, such as the eutectic mixture of potassium nitrate (40%) and sodium nitrate (60%), are stable at high temperatures but solidify readily at temperatures below 230°C. The installations that operate at high temperature use air, steam or molten salts as HTFs. Research programs are dedicated to the study of the nanoparticle-based fluids to modify the melting point and the heat transfer capacity or the creation of less corrosive eutectic salts adapted to the levels of temperature aimed by the combination of two or several salts.

Concerning the storage materials, studies have been conducted on concrete in order to evaluate the thermal stability (Laing et al. 2006; John et al. 2013; Salomoni et al. 2014). Concrete is an interesting storage material, as it is cheap, highly resistant, easy to manufacture and readily available. The first tests have shown the presence of cracks and a separation between the concrete and the tubes through which the HTF circulates. Following these tests, a large scale test module was fabricated. It underwent cyclical tests to simulate the charge and discharge at various levels of temperature between 200°C and 400°C. Additional tests have been conducted at temperatures up to 500°C for several thousand hours to study the influence of temperature on the stability of the material and the mechanical strength of concrete. The importance of concrete as sensible heat storage material up to 500°C has also been shown.

Within the European research program (DISTOR) where the (KNO_3/$NaNO_3$)/graphite composite was tested (Pincemin et al. 2008), it was shown that the use of PCMs of industrial purity required the evacuation of impurities before any thermal cycling, the risk of material deterioration being otherwise run.

The compatibility between envelope, storage materials and HTFs determines the lifetime of the system. Once the storage medium is selected, it is important to identify the materials that will be used for the remaining components of the system including the storage tank, the HTF and, if necessary, the encapsulation as well as the heat transfer intensification materials. In certain cases, for example a packed bed, the encapsulation or the confinement of the material may be unnecessary, as the storage material may be in direct contact with the transfer fluid. However, in latent heat storage systems, PCMs must be encapsulated to avoid any salting-out problem during phase change. In addition to providing sufficient transfer surface and not limiting thermal exchanges with the transfer fluid, the encapsulation material must be stable and compatible with the storage material. The encapsulation of PCMs must thus meet a certain number of technical demands such as the thermal and structural stability, mechanical strength, elasticity, resistance to corrosion and easy handling.

Corrosion is thus a key element in the selection of materials (Sequeira 2019). It reduces the life cycle of the system and modifies its performances. The data on salt corrosion at high temperature with thermal cycling are rather rare. It is important to cite Bradshaw et al. (2002) or Goods et al. (2004), who studied the corrosion of steels in contact with molten salts used at these temperatures. They highlighted the resistance of stainless steels used to fabricate the hot tank in a two-tank storage subjected to corrosion during the thermal cycles with molten nitrate salts. The tests conducted until present show that the use of carbon steel is acceptable in the operation with nitrate salts at temperatures below 400°C. For higher temperatures, it is preferable to use stainless steel for the envelope materials of tanks or other components of the system.

8.3.5. *Other components*

The other components of the storage are selected depending on the ranges of temperature, ranges of flow rates and the fluids used. Figure 8.19 illustrates the relative complexity of a two-tank thermal storage. The insulating material must be able to insulate the tank and minimize the thermal losses.

Hence, the foundation of the tank consists of a concrete slab, the external insulation is made from glass wool and the internal insulation is composed of refractory bricks.

Figure 8.19. *Two-tank thermal storage equipment (source: Aalborg). For a color version of this figure, see www.iste.co.uk/odru/heat1.zip*

HEs must be designed with temperature differences compatible with the solar field and the storage. The design must also take into account the loss of charge in the HEs. Shell and tube HEs are preferred for these applications (Figure 8.20). The higher pressure HTF circulates on the tube side (oil) and the storage fluid (molten salt) circulates on the shell side.

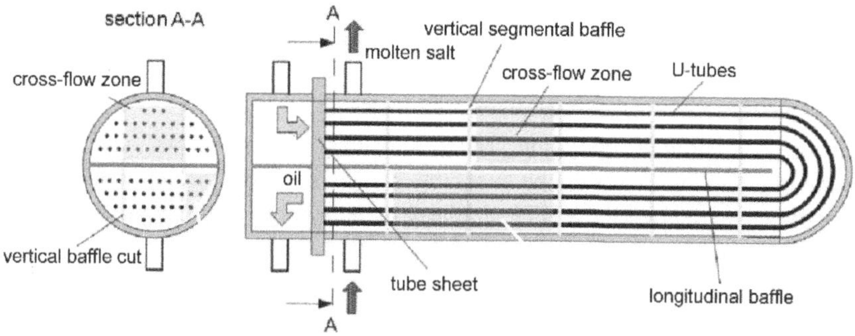

Figure 8.20. *Shell and tube heat exchanger. For a color version of this figure, see www.iste.co.uk/odru/heat1.zip*

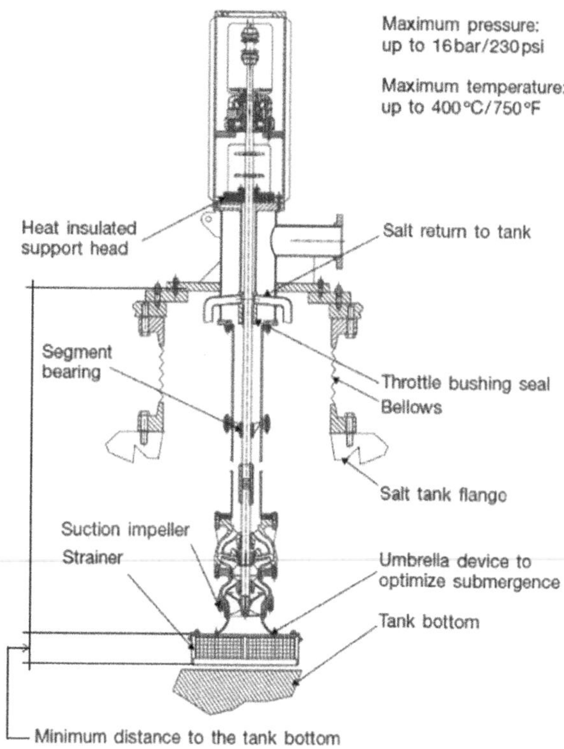

Figure 8.21. *Molten salt pump (Sulzer© VEY molten salt pump)*[1]

1. Available at: www.sulzer.com.

The pumps must be adapted to the fluid and to the working temperature level. In the case of molten salt storage, the pumps are of the type illustrated in Figure 8.21. These pumps are located at the top of the tank to carry the molten salt.

Given the variable operation, daily start-ups and shut-offs, and also the temperatures and pressures to be met, the choice is made for horizontal double suction pumps to deal with the HTF similarly to the synthetic oil (Figure 8.22).

Figure 8.22. *Horizontal pump for synthetic oil. For a color version of this figure, see www.iste.co.uk/odru/heat1.zip*

8.4. Analysis of systems

As previously noted, it is obvious that good knowledge of the efficiency of the storage system is very important for determining the overall performances of the electric power plant. The assembly of selected components plays an essential role in obtaining the expected storage density, the expected charge and discharge capacities and the considered costs. At the system level, several factors determine the thermal behavior and therefore the energy and exergy efficiency of the unit.

The key demands for the development of a thermal energy storage for the solar power plants are indicated as follows (Medrano et al. 2010; Cabeza 2015):

– high storage capacity and therefore energy density of the storage material;

– efficient heat transfer between the storage material and the transfer fluid by means of a properly designed HE;

– rapid response to changes between charge and discharge modes;

– low chemical interaction of the storage material, transfer fluid and envelope materials;

– good chemical stability of the storage material and of the fluid during the number of charge/discharge cycles corresponding to the lifetime of the power plant (30 years);

– high thermal efficiency and low parasitic electrical power for the system;

– low impact on the environment;

– low cost of the storage material, taking into account grey energy and CO_2 content;

– usability and low operating and maintenance costs;

– feasibility of the scaling of storage designs to provide electricity according to the expected cycles and a high charge rate.

In order to conduct the theoretical analysis and the modeling of the previously described storage systems and given their specificities, various cases can be identified.

8.4.1. *Active systems*

8.4.1.1. *Two-tank storage*

In most solar power plants, storage relies on the principle of two tanks with molten salt. There are systems with direct or indirect circulation.

Let us first consider the case of the system with direct circulation in which the molten salt used as storage fluid circulates between the two tanks and serves also as HTF. It is important to mention the Thémis power plant that operated between 1983 and 1986 or Gemasolar Thermosolar Plant, also known as Solar Tres, which was developed by Torresol Energy in Spain and inaugurated in 2011 (Figure 8.23).

The molten salt tank can be considered an open system (Figure 8.24) for which an energy balance is drawn.

Figure 8.23. *Storage tank, Gemasolar. For a color version of this figure, see www.iste.co.uk/odru/heat1.zip*

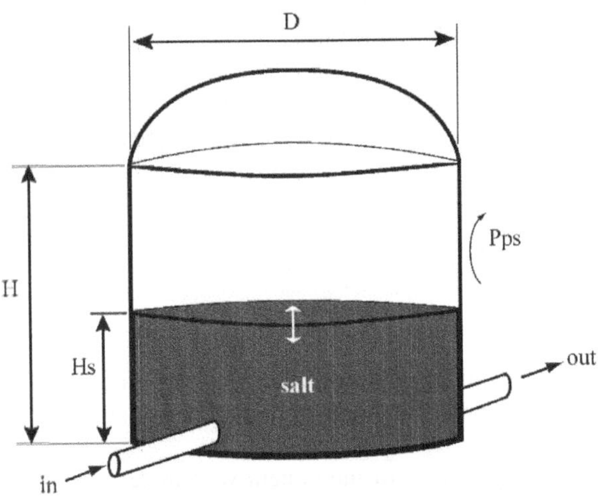

Figure 8.24. *Molten salt tank*

The internal energy variation of the salt in the tank is given by the following equation:

$$\frac{\mathrm{D}(m_s . C_{p,s} . (T_s - T_o)}{\mathrm{dt}} = \dot{m}_{s,in} . C_{p,s,in}\left(T_{s,in} - T_o\right) - \dot{m}_{s,out} . C_{p,s}(T_s - T_o)$$

$$-U_{tank}.A_{tank}.(T_s - T_{amb}) + P_{net} \qquad [8.30]$$

where:

$$m_s = m_{s,0} + (\dot{m}_{s,in} - \dot{m}_{s,out}).t \qquad [8.31]$$

and:

$$A_{tank} = \pi.D.H + \frac{\pi D^2}{4} \qquad [8.32]$$

In these equations, m_s, $\dot{m}_s$, T_s and $C_{p,s}$ are, respectively, the mass, flow rate, temperature and heat capacity of salt, D and H are the diameter and the height of the tank and T_o is the reference temperature. $m_{s,0}$ is the initial mass of salt in the tank. P_{net} is the net mechanical power essentially related to pumping. The height of salt in the tank is given by:

$$H_s = \frac{4.m_s}{\rho_S \pi.D^2} \qquad [8.33]$$

A more complete balance of the tank was proposed by Zaversky et al. (2013).

In the case of the Thémis power plant, the storage consists of about 540 tonnes of molten salt (Hitec: potassium nitrate, sodium nitrate and sodium nitrite) passing from 250 to 450°C and corresponding to a 40 MWh capacity (Figure 8.2). This choice was motivated by the willingness to fit the power plant with high-performance storage directly integrated in the primary circuit. Two pumps on the cold tank side circulate the molten salt from the cold tank to the hot tank through the solar receiver. Two other pumps fit the hot tank to circulate the salt from this tank to the cold tank by passing through the steam generator for electricity generation. The storage was dimensioned for the equivalent of 5 h of generation at 2.2 MW_e. Hence, the thermal power required for steam generation is 8.1 MW. The temperature of hot salt at the inlet of the generator is 450°C. With a 200°C drop, the variation on the enthalpy of salt is 314 kJ kg^{-1} (0.087 kWh kg^{-1}) and the required amount of hot salt is 462,000 kg. The total quantity of salt in the primary circuit is slightly higher: 537,000 kg. The salt flow rate is 26 kg s^{-1} to reach an exchanged power of 8.1 MW. The superheated steam exits at 445°C at 42 bars and a flow rate of 3.3 kg s^{-1}.

To prevent salt degradation, a gaseous sky mostly composed of nitrogen is maintained in the tanks. In this case, the choice of pumps and wall materials proves to be essential. It is important to note that the storage system as a whole (tanks, salt, hot pumps, valves, insulation, thermal tracing) represents less than 5% of the total cost of the power plant. The design developed at Thémis proves that it is possible to completely decouple the solar resource from electric power generation. This showed that storage mitigates the impact of weather fluctuations and allows a generation adapted to the conditions of the power grid.

The storage in Gemasolar Thermosolar Plant is also based on two molten salt tanks (Figure 8.25). It has a capacity of 8,500 tonnes of molten salts (nitrates) providing a 15-h autonomy, which is approximately 670 MWh_{th}. Thanks to the storage, the charge rate of the power plant reaches 55%, which is a generated power of 19.9 MW over 5,000 h annually.

Figure 8.25. *Gemasolar power plant with its two-tank storage. For a color version of this figure, see www.iste.co.uk/odru/heat1.zip*

Let us now consider the case of active systems with indirect circulation. It proves that most power plants being developed and operating, mainly in Spain and in the United States, rely on this principle. Oil is a HTF that flows through the tubes located in the focal point of the parabolic trough concentrators or at the level of the receivers of the tower power plant. An HE can be sued to transfer the heat from the oil to the molten salt that circulates between the cold tank and the hot tank during the storage phase. During the discharge phase, the molten salt passes from the hot tank to the cold tank releasing its heat to the oil that supplies the steam cycle and thus allows electricity generation.

Let us consider once again the example of the Andasol power plant. This installation with parabolic trough mirrors, with a net electric power of 49.9 MW, is fit with a two-tank storage containing 28,500 tonnes of molten salts (60% $NaNO_3$/40% KNO_3) (Figure 8.26). This 1,010 MWh_{th} storage, corresponding to 7.5 h of full power generation, provides a supply of the turbine ranging between 20.0% of its nominal power for 36 h and 90.2% for 8.3 h.

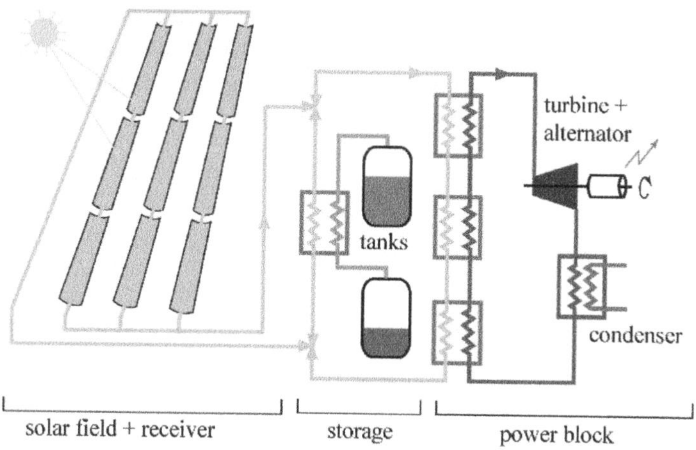

Figure 8.26. *Schematic representation of the Andasol power plant. For a color version of this figure, see www.iste.co.uk/odru/heat1.zip*

The two-tank molten salt storage exchanges heat with oil from the solar field. The charge consists of transferring to the molten salt the heat convected by the oil at 393°C through an exchanger. The salt taken in the cold tank at 292°C counterflows and is stored at 385°C. During the discharge, the system operates backwards. The transfer fluid recovers the heat released by the molten salt that flows from the hot tank to the cold tank. This heat generates the steam that operates the turbine.

The storage fluid has a solidification temperature below the lowest working temperature of the storage system and a decomposition temperature above the highest reached temperature. The solar field was oversized so that it can charge the storage system while providing thermal energy to the supply block. The number of hours provided by the storage therefore depends on the temperature range of the system, the quantity of material present in the storage and the thermal properties of the material and the energy needs of the power block.

8.4.1.2. *Storage with fluidized solid particles*

In order to improve the storage capacity of the air, it is possible to charge it with particles (Flamant et al. 2013). It is a matter of integrating a fluidized bed system in a receiver and associating it with granular material storage tanks and systems for the transportation of particles between various components. In the NextCSP project developed by the PROMES Laboratory and illustrated in Figure 8.27, the storage material is olivine. A two-tank storage design using quartz sand as storage material is studied by Warerkar et al. (2011). Sand is a low-cost storage material that contributes to the economic interest in the system. The choice of materials used to produce these particles is essentially determined by the solar receivers and the transportation of these particles in the full process (Ho 2016).

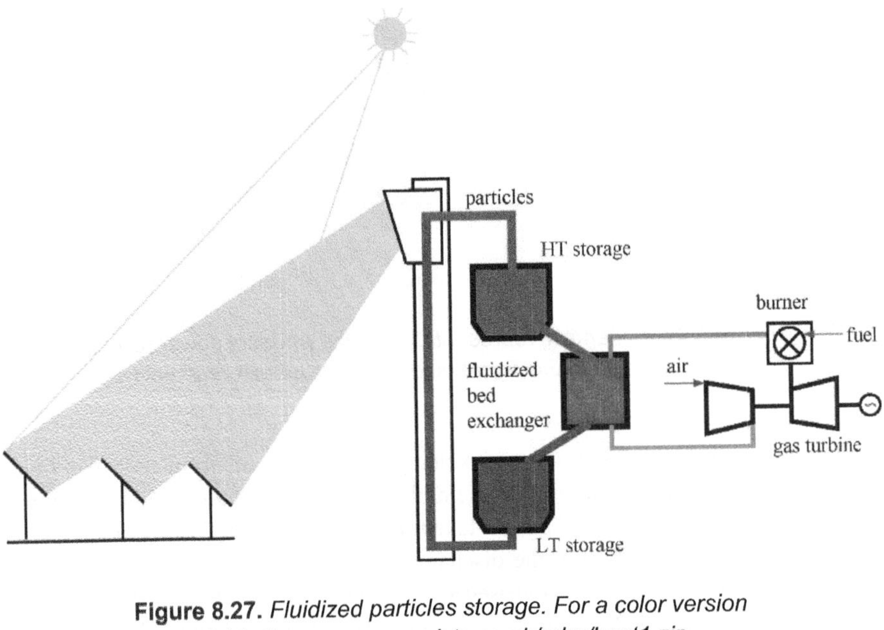

Figure 8.27. *Fluidized particles storage. For a color version of this figure, see www.iste.co.uk/odru/heat1.zip*

It is important to note that the pumping power for a grain size above 10 mm remains below 1% of the generated power. Experimental studies on the heat transfer between air and a bed of particles (iron ore, pellets of aluminum, steel or zirconium oxide, etc.) have been conducted in temperature ranges between 700 and 1,000°C, depending on tested materials. Generally speaking, the use of air charged with particles at high temperature is interesting in energy and economic terms. It is a matter of intensifying heat exchanges not only at the level of the receiver, but also of other components of the processes while mastering the transportation of particles.

8.4.1.3. *Steam accumulator*

Besides synthetic oils and molten salts in the active storage, water can be used as storage fluid in the systems known as steam accumulators. In these systems (Figure 8.28), charging occurs when superheated steam or saturated water penetrates in a storage tank under pressure that contains saturated steam and saturated water. If the system is charged with superheated steam, water temperature and pressure in the tank increases, thus modifying the saturation state of the initial mass.

On the other hand, if the saturated liquid is used to charge the system, pressure and temperature remain constant despite the volume increase of the water mass. During the discharge and therefore the generation of saturated steam, the pressure in the storage tank decreases (Figure 8.29). If superheated steam is expected, a secondary storage system is required to increase steam temperature.

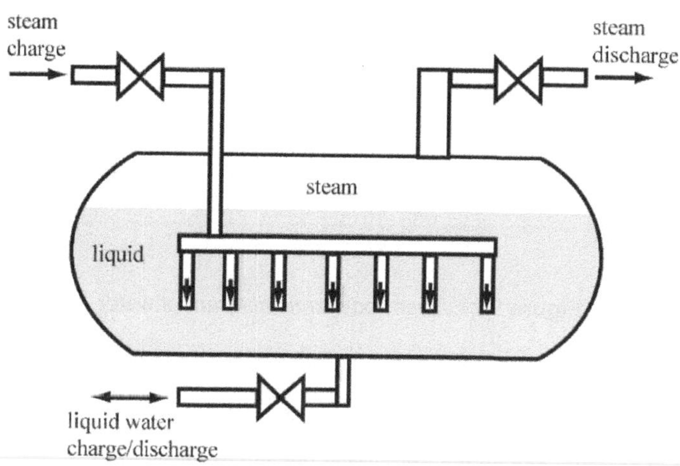

Figure 8.28. *Steam accumulator*

The model describing the phenomenon relies on several hypotheses such as the absence of heat transfer between the fluid and the tank wall, and also with the environment and the thermal equilibrium of the fluid in the tank (Steinmann and Eck 2006; Adinberg et al. 2010). During discharge, the variation of internal energy of the liquid water in the tank corresponds to the enthalpy flux of the mass of fluid (saturated steam) exiting the tank at pressure p_{res}:

$$d(m_{res}.u_{res}) = h(p_{res})dm_{res} \tag{8.34}$$

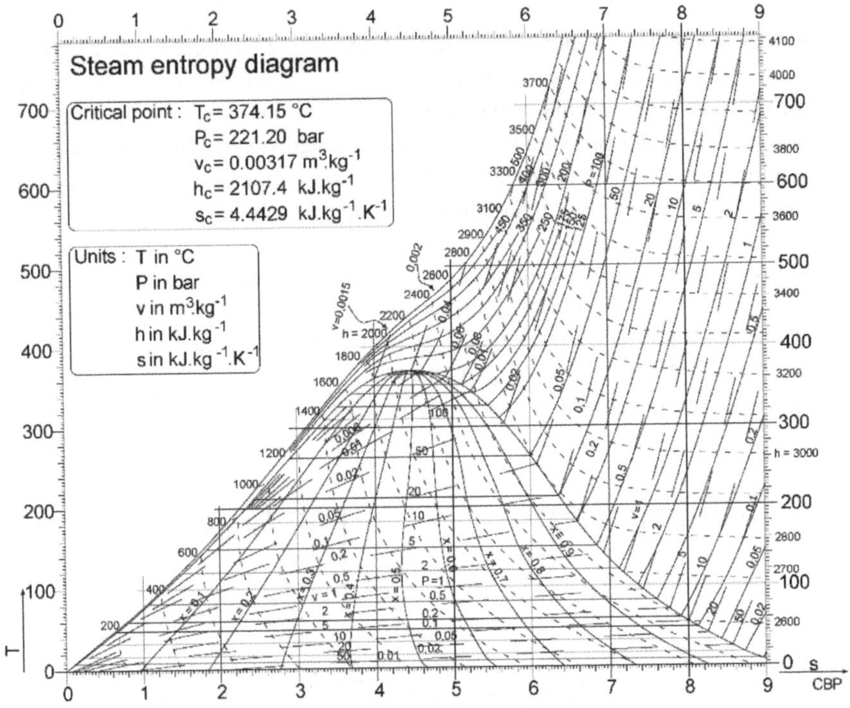

Figure 8.29. *Thermodynamic diagram of water*

The mass of saturated steam generated during discharge is obtained by integration. The quantity of saturated steam delivered by the accumulator depends on the initial pressure and on the pressure at the end of discharge.

To a first approximation, the mass of saturated steam delivered is given by:

$$m_{vs}\Delta h_{evap} = m_{liq}C_{liq}[T_{sat}(p_{init}) - T_{sat}(p_{fin})] \tag{8.35}$$

where C_{liq} is the average specific heat of saturated water. Using Watson's equation, it is possible to determine the heat of evaporation Δh_{evap} :

$$\Delta h_{evap} = \Delta h_{evap,ref} \left(\frac{1-T/647}{1-T_{ref}/647}\right)^{0.38} \quad [8.36]$$

and using Antoine's equation, the saturation temperature T_{sat} can be determined:

$$T = \frac{3985}{11.934-\ln(p_{sat})} - 39.05 \quad [8.37]$$

where temperatures are expressed in Kelvin and p_{sat} in bar.

Steam accumulators are well adapted to direct steam generation (DSG) solar power plants, in which steam is directly generated in the solar field and then used in the power block to generate energy. Steam excess, generated during the periods of strong solar radiation, can be diverted to the steam accumulator. These systems can also be used both as a storage process and phase separator in DSG installations operating in recirculation mode. In this configuration, saturated steam is separated from the saturated liquid and then sent to a superheater.

These systems require pressurized tanks, which are rather expensive. Moreover, they have low volumetric energy densities and water volumetric storage capacity ranging from 20 to 30 kWh·m^{-3}. Therefore, they are useful in buffer storage.

The Spanish power plant PS10 is an example of industrial application, counting four steam accumulators at 40 bars and 245°C for a storage capacity of 20 MWh (Figure 8.30). When the energy is required to cover a transition period, such as a cloudy spell, saturated water is recovered at variable pressure, from 40 bars to the minimal pressure authorized by the system for the operation of the turbine at partial charge of 50%. Other existing variants are proposed by Steinmann and Eck (2006) (Figures 8.31 and 8.32).

In order to increase the storage capacity, a proposal was made to place PCMs either directly inside or outside of the steam accumulator. There are also indirect storage configurations in which an exchanger allows the heat transfer between the heat transfer fluid and the steam of the accumulator.

Therefore steam accumulators represent an interesting option for short-term energy storage. The integration of steam accumulators mainly aims to compensate for the rapid transition phases of insolation. One of the main difficulties encountered in solar power plants lies in the thermomechanical constraints resulting from insolation fluctuations. The use of steam accumulators in the initial phases of operation of the power plant allows the extension of the response time of storage systems intended for operation for several hours. By improving this response time, the steam accumulators contribute to increasing the efficiency of long-term storage. Let us finally note that the investment costs of steam accumulators are dominated by the pressurized tank. The use of components such as supply water tanks or degassers as storage tanks limits the additional investment costs of steam accumulator systems.

Figure 8.30. *PS10 power plant with its steam accumulators. For a color version of this figure, see www.iste.co.uk/odru/heat1.zip*

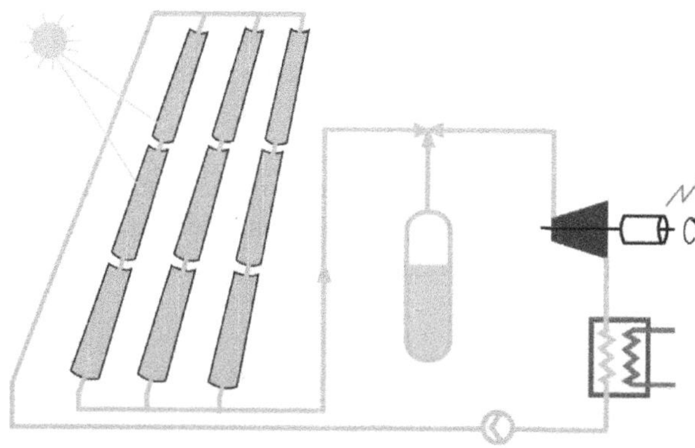

Figure 8.31. *Power plant with steam accumulator without water circulation. For a color version of this figure, see www.iste.co.uk/odru/heat1.zip*

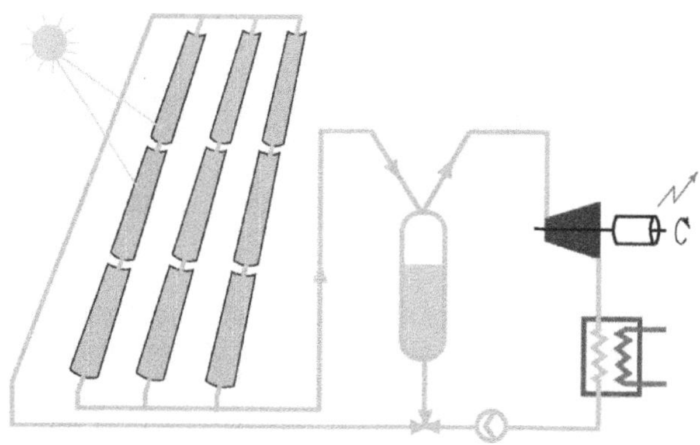

Figure 8.32. *Power plant with steam accumulator with water circulation. For a color version of this figure, see www.iste.co.uk/odru/heat1.zip*

8.4.2. *Passive systems*

8.4.2.1. *Sensible heat thermocline*

In passive storage systems, the transfer fluid carries the energy received from the solar field to the storage system during charge and receives the energy from the storage system during discharge. The hot fluid such as the oil is pumped at the top of the tank during the charge mode by progressively moving the colder liquid (Figure 8.33). A thermal gradient is created in the system and it is stabilized and preserved by buoyancy (stratification) effects. The hot fluid is at the top, while the cold fluid remains at the bottom. The high gradient thermocline zone separates the hot fluid from the cold fluid. An ideal thermocline corresponds to a piston flow. An actual thermocline spreads over a thickness that will have to be minimized to preserve the required performances.

Since the transfer fluid of power plants generates a relatively high cost, the tank is filled with cheap material. The presence of this filler material therefore reduces the quantity of fluid and acts as storage material. In this case, they are referred to as regenerators or dual media storage.

These passive storage systems can therefore use solids such as rocks, sand, concrete or materials resulting from waste.

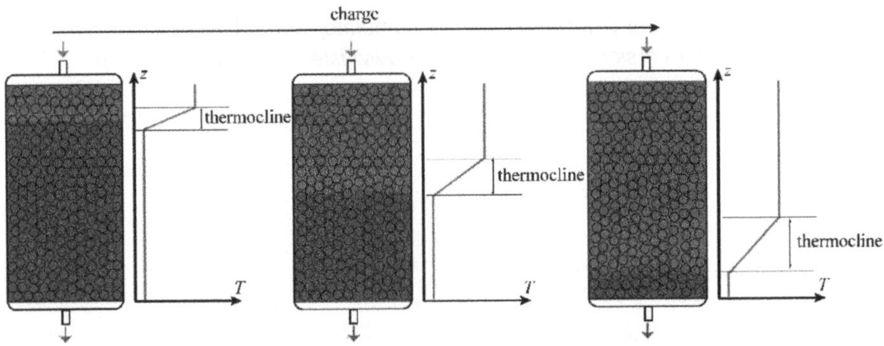

Figure 8.33. *Charge of a thermocline storage. For a color version of this figure, see www.iste.co.uk/odru/heat1.zip*

Media can be structured or unstructured. Unstructured media are typically bulk granular materials. Heat transfer may be more difficult in these passive systems than in the active systems since the storage medium is in solid phase rather than in liquid phase. Structured media may consist of stacked bricks or concrete in which metal pipes are inserted. The systems of stacked bricks draw their inspiration from the Cowper stoves used in glass furnaces. As for the concrete, Laing et al. (2006) have studied the tube exchanger integrated in concrete. The problems observed are differential expansion and degradation of contact interfaces.

The modeling of a thermocline tank relies on a system of equations with two temperatures. There are two large types of models: a 1D-1D model and a 1D-2D model. More details on the numerical differences of models are given by Esence et al. (2017). The 1D-1D model was proposed by Schumann in 1929, then resumed by Hugues et al. in 1976 and established for a bed of particles (Willmott 2001; Esence et al. 2017).

Willmott (2001) generalizes the model at various geometries and applies it more specifically to regenerators used in the industry.

The hypotheses of the model are the following:

– effective conductivity of the bed is infinite in radial direction;

– there is no temperature gradient in the particles;

– the fluid flow is unidirectional;

– the conductivity of the particle is zero in the axial direction;

– the fluid has no axial thermal dispersion;

– the properties do not depend on temperature;

– no heat loss towards the environment is considered;

– the radiative transfer between fluid and solid and between the particles in the bed is ignored.

The model is then written as follows:

– In fluid phase:

$$\varepsilon C_f \rho_f \frac{\partial T_f}{\partial t} = -u_f \varepsilon C_f \rho_f \frac{\partial T_f}{\partial x} + h_c A(T_s - T_f) \quad [8.38]$$

where A is the surface of a particle, ε is the porosity, h_c is the convective exchange coefficient and u_f is the speed of the fluid. This speed is related to the interstitial velocity v_f:

$$u_f = \frac{v_f}{\varepsilon} \quad [8.39]$$

$$v_f = \frac{\dot{m}_f}{\rho_f A_{cs}} \quad [8.40]$$

where A_{cs} is the passage section of the fluid;

– In solid phase:

$$(1 - \varepsilon) C_s \rho_s \frac{\partial T_s}{\partial t} = h_c A(T_f - T_s) \quad [8.41]$$

Let us introduce the volumetric exchange coefficient $h_v = h_c.A$. Then the number of transfer units associated with the bed of particles is:

$$NTU = \frac{h_v A_{CS} L}{\dot{m}_f C_f} \quad [8.42]$$

where L is the storage length. In case there is a temperature gradient in the particles ($Bi_D > 0{,}1$), Duffie and Beckman (2013) propose a *NTU* modified by Jefferson's correction:

$$NTU_{eff} = \frac{NTU}{1+Bi_D/5} \quad [8.43]$$

where:

$$Bi_D = \frac{h_c D}{2\lambda_s} \quad [8.44]$$

The equations can then be rewritten as follows:

$$\frac{\partial T_f}{\partial x} = \frac{NTU}{L}(T_s - T_f) - \frac{m_f}{\dot{m}_f L}\frac{\partial T_f}{\partial t} \quad [8.45]$$

$$\frac{\partial T_s}{\partial t} = NTU\frac{\dot{m}_f C_f}{m_s Cs}(T_f - T_s) \quad [8.46]$$

In order to render them dimensionless, we define:

– The dimensionless temperature:

$$T^* = \frac{T - T_{S0}}{T_{f,in} - T_{S0}} \quad [8.47]$$

– The dimensionless length:

$$\chi = L/NTU \quad [8.48]$$

– The characteristic time:

$$\tau = NTU\frac{\dot{m}_f C_f}{m_s C_s}\left[t - \frac{m_f x}{\dot{m}_f L}\right] \quad [8.49]$$

The characteristic time includes the ratio of heat stored by the solid and convected by the fluid.

The initial and boundary conditions are:

$$T_s(x,0) = T_f(x,0) = T_{s0} \quad [8.50]$$

$$T_f(0,t>0) = T_{f,in} \quad [8.51]$$

With $x^* = \frac{x}{X}$ and $t^* = \frac{t}{\tau}$, the dimensionless Schumann's model can be deduced:

$$\frac{\partial T_f^*}{\partial x^*} = T_s^* - T_f^* \quad [8.52]$$

$$\frac{\partial T_s^*}{\partial t^*} = T_f^* - T_s^* \quad [8.53]$$

with:

$$T_s^*(x^*,0) = T_f^*(x^*,0) = 0 \quad [8.54]$$

and:

$$T_{f,in}^* = 1 \quad [8.55]$$

Using the double space-time Laplace transform to solve the equations, the following relation is obtained by introducing the modified Bessel function of first type:

$$T_f^*(x^*,t^*) - T_s^*(x^*,t^*) = e^{-x^*}e^{-t^*}I_0(2\sqrt{x^*.t^*}) \quad [8.56]$$

Based on the previous equations, the expressions of dimensionless fluid and solid temperatures are deduced:

$$T_f^*(x^*,t^*) - 1 - e^{-t^*}\int_0^{x^*} e^{-\xi}I_0(2\sqrt{\xi.t^*})d\xi \quad [8.57]$$

$$T_s^*(x^*,t^*) = e^{-x^*}\int_0^{t^*} e^{-\eta}I_0(2\sqrt{x^*.\eta})d\eta \quad [8.58]$$

Moreover, according to these equations, it can be noted that:

$$T_s^*(x^*,t^*) + T_f^*(t^*,x^*) = 1 \quad [8.59]$$

A proposal was made to model the thermocline storage using transfer functions (Olivès 2021). Therefore, when a thermocline storage is subjected to variations of the input temperature of the fluid, it can be shown that it behaves as a filter, transforming the input signal into shifted output signal with a different amplitude. The thermal storage can then be considered as a set of *RC* filters in series. It transforms the input fluid temperature into a shifted and attenuated output temperature.

For a first-order filter subjected to an input step, the Laplace transform of the dimensionless fluid temperature can be written as follows:

$$\overline{T_f^*}(s) = \frac{1}{s}.\frac{1}{1+s} \quad [8.60]$$

The result of applying the inverse transform is:

$$T_f^*(\tau) = 1 - e^{-\tau} \quad [8.61]$$

For two first-order cascaded filters, we have:

$$\overline{T_f^*}(s) = \frac{1}{s}.\frac{1}{(1+s)^2} \quad [8.62]$$

hence:

$$T_f^*(\tau) = 1 - (1+\tau)e^{-\tau} \quad [8.63]$$

Finally, for n first order filters in cascade, we have:

$$\overline{T_f^*}(s) = \frac{1}{s}.\frac{1}{(1+s)^n} \quad [8.64]$$

therefore:

$$T_f^*(\tau) = 1 - e_{n-1}(\tau).e^{-\tau} \quad [8.65]$$

with:

$$e_n(\tau) = 1 + \frac{\tau}{1!} + \frac{\tau^2}{2!} + \frac{\tau^3}{3!} + \cdots + \frac{\tau^n}{n!} \quad [8.66]$$

Using the property of symmetry in space and time that can be noted for the solution of Schumann's model, the terms of the previous equation represent the behavior in time of a series of elements that constitute the storage. Hence, a filter is associated with a characteristic thickness L_u. Consequently, the order of the set of filters corresponds to the length of the storage $L = n.L_u$ (Figure 8.34).

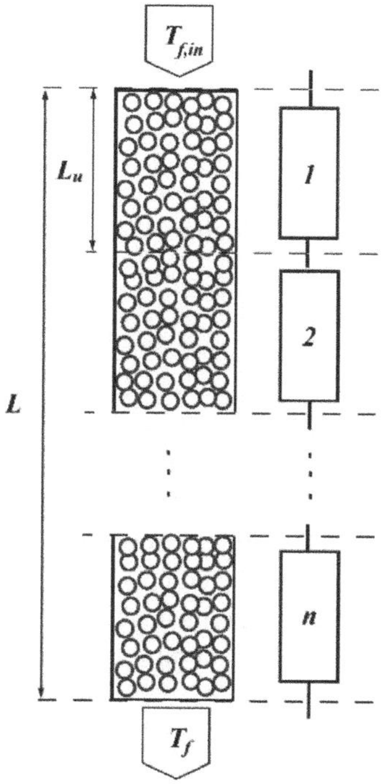

Figure 8.34. *Modeling of thermocline storage by filters in series*

The unit module serving as a base can thus be defined with a thickness L_u and an NTU_u corresponding to:

$$NTU_u = \frac{6.h.(1-\epsilon)L_u.S}{\dot{m}_f C_f D} \qquad [8.67]$$

where h is the coefficient of exchange between the fluid and the particles and S is the input section. h can be obtained using correlations such as the one of Wakao and Kaguei (Esence et al. 2017).

Using the previously employed dimensionless time that was applied to the unit module, we have:

$$\tau = NTU_u \frac{\dot{m}_f c_f}{m_{s,u} c_s} \left[t - \frac{m_f x}{\dot{m}_f L} \right] \qquad [8.68]$$

Noting that, for a gas such as the air, the second term can be ignored, the characteristic time can be expressed by:

$$\theta_o = \frac{m_{s,u} C_s}{\dot{m}_f C_f NTU_u} \qquad [8.69]$$

For a bed of particles, and considering the properties, this time can also be written as follows:

$$\theta_o = \frac{\rho_s C_s D}{6.h.(1-\epsilon)} \qquad [8.70]$$

where D is the average diameter of the particles and ϵ is the porosity of the bed.

It is important to note that θ_o does not depend on L_u. In order to account for the characteristic time θ_u of the unit module, a corrective factor related to the height L of the storage should be introduced. The comparison between the analytical solution based on the filter model and the experimental results allows the identification of this corrective factor F_c. The characteristic time of the filter θ_u is deduced, which is written as follows:

$$\theta_u = F_c \frac{\rho_s C_s D}{6.h.(1-\epsilon)} \qquad [8.71]$$

Therefore, the dimensionless temperature of the fluid at the outlet of the storage, itself subjected to a temperature step at the inlet, is given by the following relation:

$$T_f^*(t/\theta_u) = 1 - e_{n-1}(t/\theta_u).e^{-t/\theta_u} \qquad [8.72]$$

Let us consider once again the dimensional model of Schumann. It can be completed by a term describing axial diffusion. Therefore:

$$\varepsilon C_{pf} \rho_f \frac{\partial T_f}{\partial t} = \lambda_{eff,f} \frac{\partial^2 T_f}{\partial x^2} - u_f \varepsilon C_{pf} \rho_f \frac{\partial T_f}{\partial x} + h_c A(T_s - T_f) \qquad [8.73]$$

and:

$$(1-\varepsilon) C_{ps} \rho_s \frac{\partial T_s}{\partial t} = \lambda_{eff,s} \frac{\partial^2 T_s}{\partial x^2} + h_c A(T_f - T_s) \qquad [8.74]$$

where $\lambda_{eff,f}$ and $\lambda_{eff,s}$ are the effective conductivities of the fluid and of the solid, respectively.

In this case, the solution is numerical (Esence et al. 2017). On wide ranges of temperatures, it may be useful to consider the dependence of thermophysical properties on temperature.

If, from now on, the Biot number is too high, Jefferson's correction cannot be applied. It is then necessary to consider the radial dispersion in the solid; the model becomes 1D fluid and 2D solid.

It is finally possible to add a third equation relative to the wall coupled with the two others by an exchange term. This equation is all the more important as the dimensions of the system are smaller (Hoffmann et al. 2016).

The temperature profiles illustrated in Figure 8.35 are obtained.

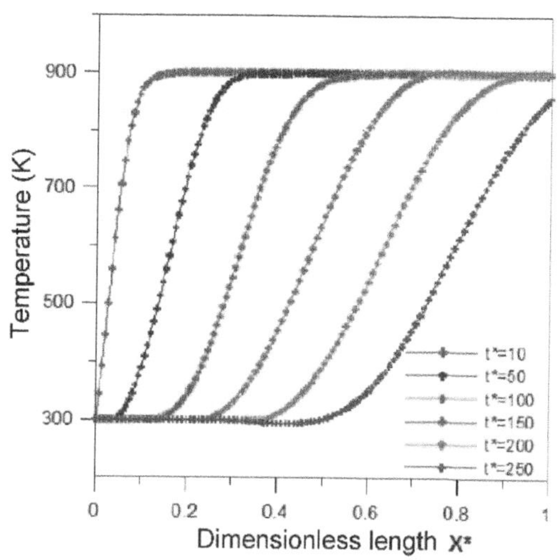

Figure 8.35. *Temperature profiles in a thermocline with various dimensionless times. For a color version of this figure, see www.iste.co.uk/odru/heat1.zip*

The filler material contributes to maintaining the gradient by reducing natural convection in the liquid phase. The objective is to have the lowest possible effective axial conductivity of the porous medium in order to avoid heat diffusion and therefore the increase in thermocline thickness. In order to improve the transfers between fluid and solid, it is important to increase the exchange surface by reducing the diameter of the particles. Nevertheless, this leads to a reduction in the passage section of the fluid and therefore an increase in charge losses. It is therefore important to reach a compromise between thermal transfers and charge losses.

Stratification is easier to preserve than in liquid alone. Moreover, it can be noted that a bed of particles stabilizes the flow and tends to homogenize the temperature in the flow section after the equivalent of several diameters of particles.

Thanks to these models, the thermal behavior of a thermocline storage is now well known. It is thus admitted that performances tend to increase with the height of the tank, which is however limited by civil engineering constraints. The Biot number associating the diameter of particles and their conductivity is also an important parameter, as it determines the thermal transfers during storage or discharge phases. A low Biot number leads to high thermal performances.

Moreover, in a solar power plant, storage must also meet other criteria related to the solar resource during charge and to the supply block during discharge. Threshold working temperatures are defined: maximal permissible temperature at tank outlet during charge and minimal permissible temperature at tank outlet during discharge. The choice of these threshold temperatures will directly influence storage efficiency (Fasquelle et al. 2018) (Figure 8.36). This storage efficiency is itself defined by the quantity of energy actually used and therefore extracted from the storage with respect to the maximal storable quantity of energy. The latter corresponds to the energy that the tank could store if it operated entirely between the low temperature and the high temperature.

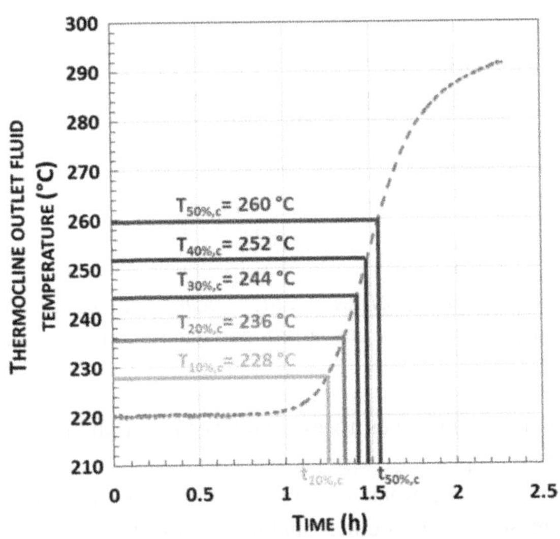

Figure 8.36. *Thermocline outlet fluid temperature of MicroSol-R (PROMES) power plant during standard charge (source: Fasquelle et al. 2018). For a color version of this figure, see www.iste.co.uk/odru/heat1.zip*

Each threshold temperature has a corresponding charge duration reflected by a different charge efficiency. This efficiency increases with the increase in temperature threshold.

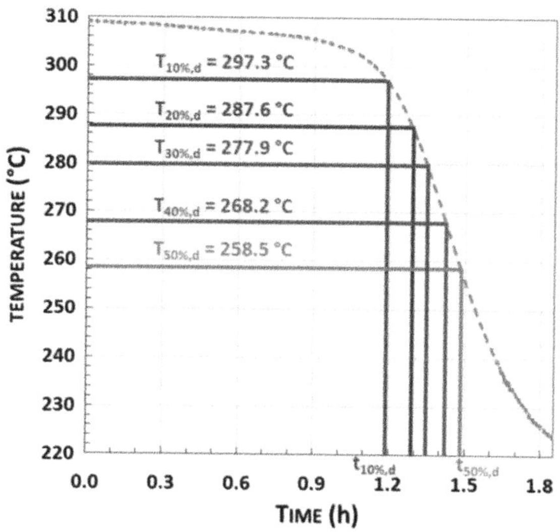

Figure 8.37. *Output temperature of the thermocline tank of MicroSol-R (PROMES) power plant during standard discharge (source: Fasquelle et al. 2018). For a color version of this figure, see www.iste.co.uk/odru/heat1.zip*

During discharge, the efficiency will be all the higher as the chosen threshold temperature will be lower (Figure 8.37). In fact, the minimal permissible temperature is imposed by the steam generator or the turbine power. Fasquelle et al. (2018) introduce dynamic threshold temperatures in order to optimize power plant operation.

The first example of a power plant with a thermocline storage was Solar One (Figure 8.38). The storage consisted of 6,170 tonnes of rock and sand, through which flowed Caloria HT43 oil between 302 and 218°C, contained in a tank with a diameter of 18.3 m and height of 13.7 m. The glass fiber insulation was 305 mm thick on the sidewall and 610 mm at the top. It was completed by 610 mm of low-density concrete on the bottom of the tank to separate the storage material from the foundation. This insulation contributed to reducing thermal losses to 3 MWh in 20 h, or less than 2% of the storage capacity that was 170 MWh.

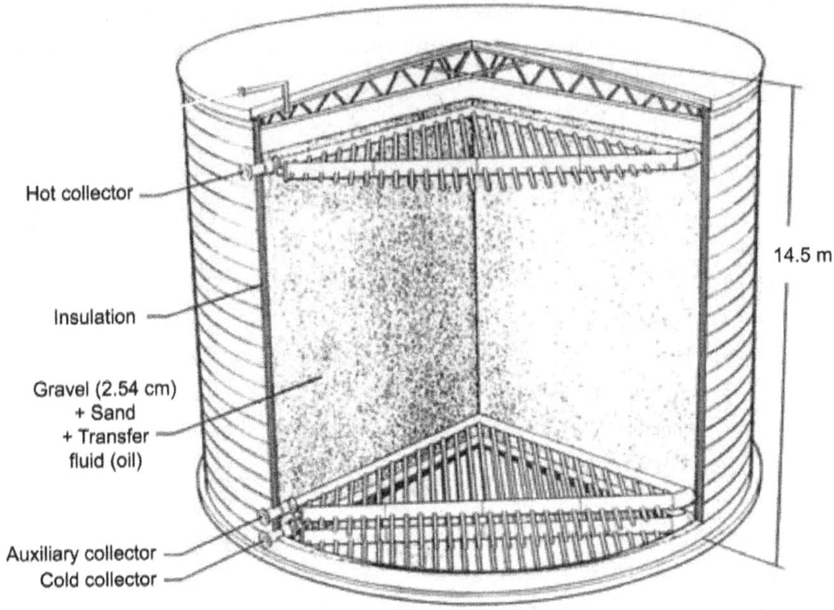

Figure 8.38. *Thermal storage with mixed filler materials of Solar One power plant, Barstow, USA*[2]

The energy used by the pumps was estimated at about 3.4 kWh/MWh$_{th}$. The storage allowed the generation of 7 MW$_e$ electricity for 4 h. The cost of heat storage reached two-thirds of the cost of the system with two tanks, both using the same molten salt. Moreover, it was noted that the presence of aluminosilicate from rocks and sand accelerated, by catalysis, the thermal cracking of oil. The operation of Solar One power plant was discontinued in 1986 due to the accidental admission of water into the tank, which led to a sudden increase in pressure and damaged the tank.

Currently, the Airlight Energy Ait-Baha pilot power plant near Agadir in Morocco couples a field of parabolic trough concentrators with an opening surface of 6,159 m^2 with a 3 MW turbine (organic Rankine cycle). Set into operation in 2014, it uses the air passing from 270 to 570°C as transfer fluid and a rock bed storage that provides the equivalent of 5 h of storage.

2. Available at: http://www.powerfromthesun.net/.

8.4.2.2. *Latent heat storage*

In order to increase energy density, the use of solid/liquid PCMs was taken into consideration. As already noted, these materials require encapsulation. A storage system design, known as RHTS (Reflux Heat Transfer Storage), was developed for solar power plants where the temperature of synthetic oil heated by the sun may reach approximately 400°C and is presented in Figure 8.39 (Laing et al. 2013). It relies on the phase change coupled with melting and vaporization that take place simultaneously in the storage system. During laboratory tests, the overall thermal conductivity of an RHTS system was determined as approximately 500 W·m^{-1}·K^{-1}, which is comparable to thermosiphon performances.

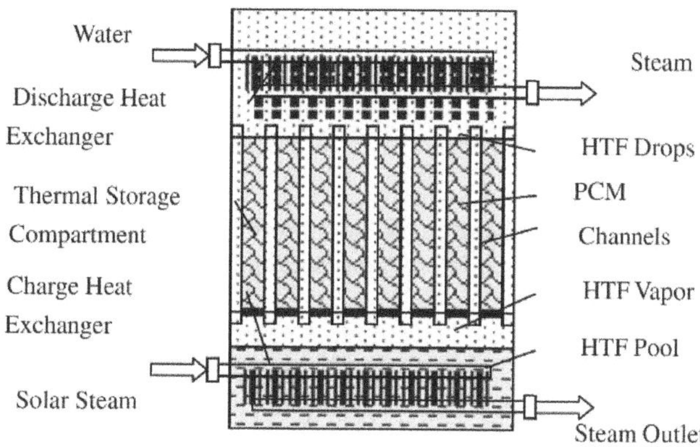

Figure 8.39. *RHTS design (source: Laing et al. 2013)*

This RHTS storage is intended to generate steam at high temperature when the generation of solar steam is temporarily interrupted. The HTF is used to transfer heat between the PCM and two HEs located outside PCM at the bottom and top of the storage. The upper HE, which is the steam generator, is supplied with high-pressure water (return condensate) to generate superheated steam during the discharge phase. Lower HE serves to charge the thermal storage. It is immersed in HTF liquid and connected to the solar field fluid. During thermal charge cycles, given HTF boiling, there is an intense flux of steam through the transport channels distributed in the PCM. By condensation on the channel surfaces and directly on PCM surface, the latent heat of steam is transferred through the walls to the PCM. During the discharge, the direction of the heat flux is reversed. Hot PCM causes the evaporation of HTF liquid and steam transmits the heat to the steam generator located at the top via condensation.

A three-part storage system illustrated in Figure 8.40 is proposed by Laing et al. (2013) for a DSG power plant, where a PCM is associated with evaporation, while the sensible heat storage is used to store the heat during water preheating and steam superheating.

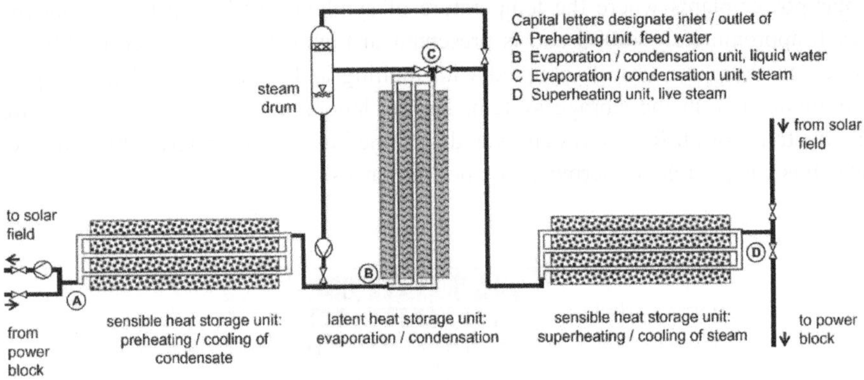

Figure 8.40. *Storage for DSG power plant combining sensible heat and latent heat (source: Laing et al. 2013)*

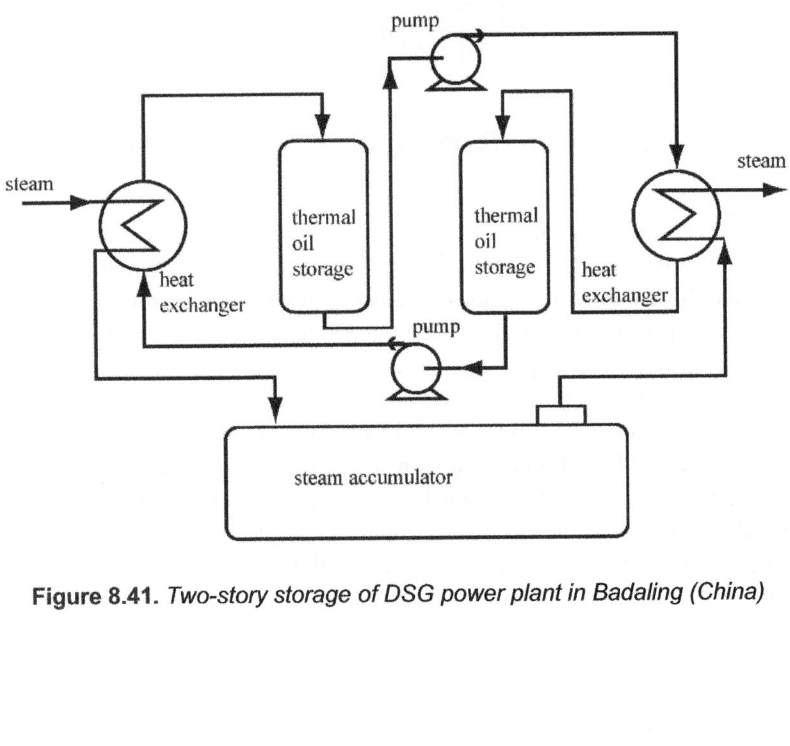

Figure 8.41. *Two-story storage of DSG power plant in Badaling (China)*

A coupled sensible/latent/sensible storage prototype, with 1 MWh_{th} capacity, was installed in the power plant at Carboneras, in Spain, in 2009. The latent heat storage module contains 14 tonnes of $NaNO_3$. It is the first demonstration of such a storage system. Cycling tests for each separate storage unit and tests of coupled systems have shown they are of technical interest. Various operational modes for PCM storage have been evaluated.

A two-story storage system was used in the 1 MW_e DSG power plant in Badaling, in China. The low temperature story is composed of a steam accumulator, which stores saturated steam at 2.35 MPa and 220.7°C (Figure 8.41). During discharge, saturated steam is converted into superheated steam via a two-tank indirect storage system that uses oil as storage fluid (Xu et al. 2012).

8.4.3. *Technical-economic analysis*

The various services brought by the storage need to be evaluated in economic terms. In order to determine the economic interest of the thermal storage, the LCOE (Levelized Cost of Electricity or the updated cost of electricity) is calculated and compared to the cost of electricity on the grid.

Then, the influence on LCOE of the various parameters involved in the storage should be analyzed in order to determine the potential gains brought by replacing one technology with another, for example. It is also possible to analyze the specific storage cost.

As an illustration, 7 h of daily storage can increase the capacity factor on average by 25–28% and can also reach 43%. Such a storage system intended for this usage requires a solar field that is necessarily larger than that for power plants without storage (Winter et al. 1991). Even though the investment is 20–40% higher, the updated annual cost of the generated electricity can be reduced by 10%–15% because of the higher charge rate and the smaller number of startups of the power block.

Quite often in solar power plants, the largest part of the cost is due to the storage material (Figure 8.42). In the case of high-temperature storage systems, many inorganic salts are proposed. Unlike the metals and oil products sold on international markets, industrial salts are rather produced at the regional level and the manufacturers are directly competing to set the price. Prices vary also from one salt to another and depending on the purity level for one salt (Table 8.6).

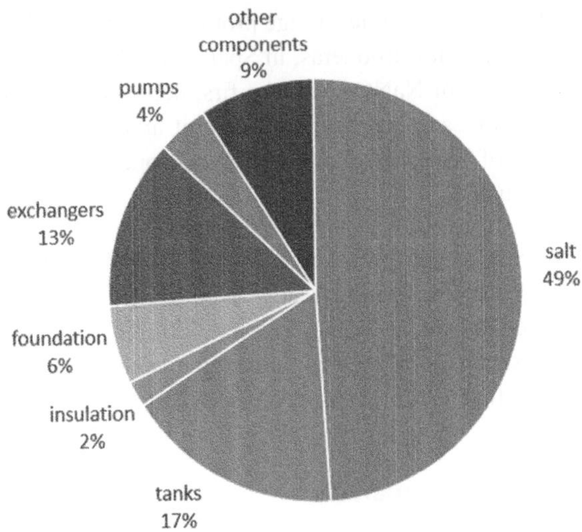

Figure 8.42. *Distribution of two-tank storage costs (source: Kuravi et al. 2013). For a color version of this figure, see www.iste.co.uk/odru/heat1.zip*

Storage material	T (°C)	ρ (kg/m3)	Cp (kJ/kg K)	λ (W/m K)	Cost (€/kg)
			Solid		
Rock	<1,000	2,200–2,600	0.6–0.7	2–5	0.1
Concrete	<500	2,250	2.0	1.3	0.7–3.0
Refractory silica brick	<700	1,820	1.0	1.5	1.0
Refractory magnesia brick	<1,200	3,000	1.15	5.0	2.0
Cofalit	<1,000	3,120	0.9	2.0	0.1
			Liquid		
Solar Salt™	290–593	1,790	1.5	0.5	0.56
Nitrate salts	265–565	1,870	1.6	0.52	0.6

Table 8.6. *Costs of materials for the storage system (according to Heller et al. (2013))*

It is important to note the work of Khamlich et al. (2021), which presents a technical and economic evaluation of five technologies for the storage of thermal energy when the production of CSP plants is integrated into a SPOT market of electricity. It is worth noting that SPOT markets generally offer interesting opportunities for higher profits by proposing prices that are more attractive than flat rates. In this context, the economic profitability is evaluated by using the net present value NPV, which appears to be a relevant indicator. Indeed, a specificity of NPV is that it can take into account the variability of prices in time.

Two-tank indirect sensible heat storage technology has the smallest financial benefits with a negative NPV because it uses expensive transfer fluids. On the other hand, two-tank direct sensible heat storage has a better profitability compared to the first technology. The latter does not include additional HE, additional valves and pumps, but uses a larger quantity of transfer fluid. Thermocline storage with a single tank and latent heat storage have the highest economic profitability among the considered technologies, though they are not yet commercialized. The latent heat storage system has however a relatively lower profitability than the thermocline tank due to the cost of PCMs. On the other hand, thermochemical heat storage has the highest investment cost and is the less profitable investment. In order to reduce the storage cost, an interesting path is the use of materials resulting from waste recycling or industrial co-products, as previously mentioned. In the same line of thought, bulk granular materials seem to be more appropriate than self-supported materials that are necessarily more expensive and involve a more complex, and consequently more costly filler phase.

The current commercial systems for the storage of thermal energy used in the solar power plants are the steam accumulators and molten salt tanks (Gonzalez-Roubaud et al. 2017). Storage represents 9–18% of the investment required for solar power plants with parabolic trough concentrators or for solar tower power plants, respectively. Therefore, the impact of storage cost reductions on LCOE cannot be ignored.

8.4.4. *Life cycle analysis, efficiency, concentration and eco-design*

Life cycle analysis (LCA) is a technique for evaluating the environmental impacts associated with all the life stages of a product (extraction of raw material, processing of materials, manufacturing, distribution, use, maintenance, disposal or recycling). It is therefore a decision-making tool for choosing a technology based on various environmental indicators.

This methodology is used to analyze the impacts of thermal energy storage systems during their lifetime. Oró et al. (2012) present the LCA for three storage systems for solar power plants: sensible heat storage on solid material (concrete), two-tank molten salt sensible heat storage and latent heat storage (Bayón et al. 2010). The authors compare the energy economies and their respective environmental impacts. Several hypothetical scenarios are studied using LCA methodology in order to highlight the differences between storage systems. The currently used molten salt two-tank storage proves to have the highest environmental impact compared to the other two storages. This stronger impact is mainly due to the complexity of the system and to the fact that it uses more storage material. Finally, due to its simplicity, the system based on solid materials has the lowest environmental impact per kWh. Figure 8.43 illustrates the difference in terms of greenhouse gases between the molten salt two-tank storage and the thermocline storage on solid material.

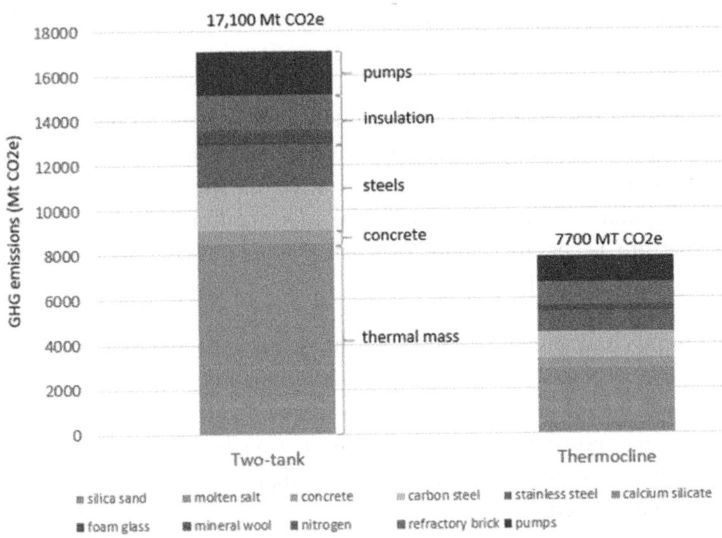

Figure 8.43. *Greenhouse gas emissions of two types of thermal storage: two molten salt/thermocline tanks on solid material (source: Sandia Nat. Lab.). For a color version of this figure, see www.iste.co.uk/odru/heat1.zip*

Burkhardt et al. (2011) evaluated the environmental impacts of a 103 MW power plant with parabolic trough concentrators using a dry or wet cooling system in the southwestern United States with 6.3 h of storage. The life cycle study leads to an estimation of greenhouse gas emissions (GHG), cumulated energy demand and the water consumption associated with manufacturing, construction, operation, dismantling and disposal of the power plant. It is important to note that the storage may represent up to approximately 40% of GHG emissions depending on the technology employed.

GHG emissions and NER (Net Energy Ratio) have been used to compare storage technologies (sensible heat, latent heat and thermochemical storage). A comparison of these technologies has been conducted by Thaker et al. (2019). By setting a common study perimeter, the analysis concerned the needs in materials and in energy, as well as the GHG emissions associated with the extraction of resources, the manufacturing of materials, transportation, operation, dismantling and disposal. The average values of GHG emissions for the two-tank direct sensible heat storage and for the one tank direct sensible heat storage are 15 and 11 gCO_2-eq/kWh, which is lower than for other systems. This can be explained by the fact that GHG emissions, due to the production of HTF in the case of two-tank direct sensible heat storage and one-tank direct sensible heat storage, represent, respectively, about 47% and 36% of GHG emissions of the life cycle, which is below those of two-tank indirect sensible heat storage (63%) and of latent heat storage (70%). As for the thermochemical storage, it uses ammonia as HTF. Since ammonia production is highly energy intensive, the GHG emissions related to ammonia production represent about 67% of life cycle emissions. We should therefore ideally consider the system that requires fewer materials during the manufacturing phase, as it will significantly reduce GHG emissions and will improve NER. For these reasons, the two-tank direct sensible heat storage and the one tank direct sensible heat storage can be considered more environmentally friendly.

Referring more specifically to the storage material, life cycle analyses have been conducted. The objective is to identify the relative impact of various materials. Nahhas et al. (2020) compare various technologies and associated materials: "two molten salt tanks", "thermocline/molten salts with recycled ceramics", "thermocline/molten salts with mullite" and "thermocline/air with basalt". The authors show a significant reduction of environmental impact. The latter were reduced by 65% for primary energy demand, 60% for potential climate change and 80% for water consumption compared to the standard two-tank molten salt storage system. The energy payback times for basalt/air storage technology are very short (about 2 months) for an expected lifetime of over 25 years.

Lalau et al. (2016) quantify the environmental footprint of recycled ceramics obtained from industrial waste by conducting a comparative analysis of the life cycle. Compared to the two-tank molten salt technology-based storage, the environmental impact of the storage unit using recycled ceramics resulting from industrial waste is reduced by 40% in terms of climate change potential, 30% in primary energy demand and 60% in water consumption. This impact is calculated for a wide range of recycled ceramics favoring the use of secondary raw materials for which the inerting process is upstream of the life cycle. If included, the energy payback time of the storage remains less than 3 years for an expected operational life of approximately 25–30 years. Such a short return on investment time is a significant advantage for further promoting the higher added value use and recycling.

Studies on the eco-design of thermal storage have been published very recently (Le Roux et al. 2021). They rely on multicriteria optimization that associates the energy analysis, environmental impact via a LCA and the economic analysis. They guide the choices toward energy-efficient, economically profitable and environmentally viable solutions.

In conclusion, there has been a strong development of CSP plants over the past two decades. In 10 years, the cost of electricity generated by solar power plants was reduced by a factor of five, because of the development and construction of demonstrative installations and pilot plants (Prieto et al. 2020). Solar power plants contribute to mitigating climate change by decarbonizing our energy system. The cheap and large-scale thermal storage integrated to these power plants allows a grid-fit power generation. It is an economically interesting alternative, which ensures a flexible and secure supply of solar power.

8.5. References

Achkari, O. and El Fadar, A. (2020). Latest developments on TES and CSP technologies: Energy and environmental issues, applications and research trends. *Applied Thermal Engineering*, 167, 114806.

Adinberg, R., Zvegilsky, D., Epstein, M. (2010). Heat transfer efficient thermal energy storage for steam generation. *Energy Conversion and Management*, 51(1), 9–15.

Bayón, R., Rojas, E., Valenzuela, L., Zarza, E., León, J. (2010). Analysis of the experimental behaviour of a 100 kWth latent heat storage system for direct steam generation in solar thermal power plants. *Applied Thermal Engineering*, 30(17/18), 2643–2651.

Bradshaw, R.W., Dawson, D.B., De la Rosa, W., Gilbert, R., Goods, S.H., Hale, M.J., Jacobs, P., Jones, S.A., Kolb, G.J., Pacheco, J.E. et al. (2002). Final test and evaluation results from the solar two project. *Contract*, 294, 55–63.

Burkhardt III, J.J., Heath, G.A., Turchi, C.S. (2011). Life cycle assessment of a parabolic trough concentrating solar power plant and the impacts of key design alternatives. *Environ. Sci. Technol.*, 45, 6.

Cabeza, L.F. (2015). *Advances in Thermal Energy Storage Systems*. Elsevier, Duxford.

Calvet, N., Gomez, J. C., Faik, A., Roddatis, V.V., Meffre, A., Glatzmaier, G.C., Doppiu, S., Py, X. (2013). Compatibility of a post-industrial ceramic with nitrate molten salts for use as filler material in a thermocline storage system. *Applied Energy*, 109, 387–393.

Duffie, J.A. and Beckman, W.A. (2013). *Solar Engineering of Thermal Processes*. John Wiley & Sons, Hoboken.

Esence, T., Bruch, A., Molina, S., Stutz, B., Fourmigué, J.-F. (2017). A review on experience feedback and numerical modeling of packed-bed thermal energy storage systems. *Solar Energy*, 153, 628–654.

Faas, S.E., Thorne, L.R., Fuchs, E.A., Gilbertsen, N.D. (1986). 10 MWe/solar thermal central receiver pilot plant: Thermal storage subsystem evaluation. Final report, 113.

Faik, A., Guillot, S., Lambert, J., Véron, E., Ory, S., Bessada, C., Echegut, P., Py, X. (2012). Thermal storage material from inertized wastes: Evolution of structural and radiative properties with temperature. *Solar Energy*, 86(1), 139–146.

Fasquelle, T., Falcoz, Q., Neveu, P., Hoffmann, J.-F. (2018). A temperature threshold evaluation for thermocline energy storage in concentrated solar power plants. *Applied Energy*, 212, 1153–1164.

Flamant, G., Gauthier, D., Benoit, H., Sans, J.L., Garcia, R., Boissière, B. (2013). Dense suspension of solid particles as a new heat transfer fluid for concentrated solar thermal plants: On-sun proof of concept. *Chem. Eng. Sci.*, 102, 567–576.

Friefeld, J.M., Moore, A.E., Morgan, G.R. (1982). Design and construction of the "Solar One" thermal energy subsystem. In *17th IECEC – IEEE*. Proceedings of the Seventeenth Intersociety Energy Conversion Engineering Conference, Los Angeles, 1468–1473.

Galione, P.A., Pérez-Segarra, C.D., Rodríguez, I., Oliva, A., Rigola, J. (2015). Multi-layered solid-PCM thermocline thermal storage concept for CSP plants. Numerical analysis and perspectives. *Applied Energy*, 142, 337–351.

Gil, A., Medrano, M., Martorell, I., Lázaro, A., Dolado, P., Zalba, B., Cabeza, L.F. (2010). State of the art on high temperature thermal energy storage for power generation. Part 1 – Concepts, materials and modellization. *Renewable and Sustainable Energy Reviews*, 14(1), 31–55.

González-Roubaud, E., Pérez-Osorio, D., Prieto, C. (2017). Review of commercial thermal energy storage in concentrated solar powerplants: Steam vs. molten salts. *Renewable and Sustainable Energy Reviews*, 80(C), 133–148.

Goods, S.H. and Bradshaw, R.W. (2004). Corrosion of stainless steels and carbon steel by molten mixtures of commercial nitrate salts. *J. Mater. Eng. Performance*, 13, 78–87.

Gutierrez, A., Miró, L., Gil, A., Rodríguez-Aseguinolaza, J., Barreneche, C., Calvet, N., Py, X., Fernández, A.I., Grágeda, M., Ushak, S. et al. (2016). Advances in the valorization of waste and by-product materials as thermal energy storage (TES) materials. *Renewable and Sustainable Energy Reviews*, 59, 763–783.

Hänchen, M., Brückner, S., Steinfeld, A. (2011). High-temperature thermal storage using a packed bed of rocks: Heat transfer analysis and experimental validation. *Applied Thermal Engineering*, 31, 1798–1806.

Heller, L. (2013). Literature review on heat transfer fluids and thermal energy storage systems in CSP plants. Solar Thermal Energy Research Group Report, Stellenbosch University.

Ho, C.K. (2016). A review of high-temperature particle receivers for concentrating solar power. *Applied Thermal Engineering*, 109, 958–969.

Hoffmann, J.-F., Fasquelle, T., Goetz, V., Py, X. (2016). A thermocline thermal energy storage system with filler materials for concentrated solar power plants: Experimental data and numerical model sensitivity to different experimental tank scales. *Appl. Therm. Eng.*, 100, 753–761.

Hoffmann, J.-F., Vaitilingom, G., Henry, J.F., Chirtoc, M., Olives, R., Goetz, V., Py, X. (2018). Temperature dependence of thermophysical and rheological properties of seven vegetable oils in view of their use as heat transfer fluids in concentrated solar plants. *Solar Energy Materials and Solar Cells*, 178, 129–138.

Hoffmann, J.-F., Fasquelle, T., Vaitilingom, G., Olives, R., Py, X., Goetz V. (2019). Compatibility of vegetable oils with solid filler materials for thermocline thermal energy storage systems. *Solar Energy Materials and Solar Cells*, 200, 109932.

John, E., Hale, M., Selvam, P. (2013). Concrete as a thermal energy storage medium for thermocline solar energy storage systems. *Solar Energy*, 96, 194–204.

Khamlich, I., Zeng, K., Flamant, G., Baeyens, J., Zou, C., Li, J., Yang, X., He, X., Liu, Q., Yang, H. et al. (2021). Technical and economic assessment of ther-mal energy storage in concentrated solar power plants within a spot electricity market. *Renewable and Sustainable Energy Reviews*, 139, 110583.

Khare, S., Dell'Amico, M., Knight, C., McGarry, S. (2013). Selection of materials for high temperature sensible energy storage. *Solar Energy Materials and Solar Cells*, 115, 114–122.

Kuravi, S., Trahan, J., Goswami, D.Y., Rahman, M.M., Stefanakos, E.K. (2013). Thermal energy storage technologies and systems for concentrating solar power plants. *Progress in Energy and Combustion Science*, 39(4), 285–319.

Laing, D., Steinmann, W.D., Tamme, R., Richter, C. (2006). Solid media thermal storage for parabolic trough power plants. *Solar Energy*, 80(10), 1283–1289.

Laing, D., Bauer, T., Breidenbach, N., Hachmann, B., Johnson, M. (2013). Development of high temperature phase-change-material storages. *Applied Energy*, 109, 497–504.

Lalau, Y., Py, X., Meffre, A., Olives, R. (2016). Comparative LCA between current and alternative waste-based TES for CSP. *Waste and Biomass Valorization*, 7, 1509–1519.

Le Roux, D., Lalau, Y., Rebouillat, B., Neveu, P., Olivès, R. (2021). Thermocline thermal energy storage optimisation combining exergy and life cycle assessment. *Energy Conversion and Management*, 248, 114787.

Medrano, M., Gil, A., Martorell, I., Potau, X., Cabeza, L.F. (2010). State of the art on high-temperature thermal energy storage for power generation. Part 2 – Case studies. *Renewable and Sustainable Energy Reviews*, 14(1), 56–72.

Mehling, H. and Cabeza, L.F. (2008). *Heat and Cold Storage with PCM: An Up to Date Introduction into Basics and Applications*. Springer, New York.

Meyer, O.A.J., Dinter, F., Govender, S. (2019). Optimisation in operating strategies for concentrating solar power plants. *Renewable Energy Focus*, 30, 78–91.

Montes, M.J., Abánades, A., Martínez-Val, J.M., Valdés, M. (2009). Solar multiple optimization for a solar-only thermal power plant, using oil as heat transfer fluid in the parabolic trough collectors. *Solar Energy*, 83(12), 2165–2176.

Nahhas, T., Py, X., Olives, R. (2020). Life cycle assessment of air-rock packed bed storage system and its comparison with other available storage technologies for concentrating solar power plants. *Waste and Biomass Valorization*, 11, 5.

Olivès, R. and Touzo, A. (2021). Modélisation analytique d'un stockage thermocline pour la récupération de chaleur fatale ou pour centrale solaire à concentration. In *Congrès SFT2021*, Belfort.

Oró, E., Gil, A., Gracia, A., Boer, D.R., Cabeza, L.F. (2012). Comparative life cycle assessment of thermal energy storage systems for solar power plants. *Renewable Energy*, 44, 166–173.

Pardo, P., Deydier, A., Anxionnaz-Minvielle, Z., Rougé, S., Cabassud, M., Cognet, P. (2014). A review on high temperature thermochemical heat energy storage. *Renewable and Sustainable Energy Reviews*, 32, 591–610.

Pincemin, S., Olives, R., Py, X., Christ, M. (2008). Highly conductive composites made of phase change materials and graphite for thermal storage. *Sol. Energy Mater. Sol. Cell*, 92(6), 603–613.

Prieto, C., Fereres, S. Cabeza, L.F. (2020). The role of innovation in industry product deployment: Developing thermal energy storage for concentrated solar power. *MDPI, Energies*, 13(2943), 1–19.

Py, X., Olives, R., Mauran, S. (2001). Paraffin/porous-graphite-matrix composite as a high and constant power thermal storage material. *Int. J. Heat. Mass. Transf.*, 44, 2727–2737.

Py, X., Calvet, N., Olives, R., Meffre, A., Echegut, P., Bessada, C., Veron, E., Ory, S. (2011). Recycled material for sensible heat based thermal energy storage to be used in concentrated solar thermal power plants. *J. Sol. Energy Eng.*, 133(3), 031008–031015.

Py, X., Azoumah, Y., Olives, R. (2013). Concentrated solar power: Current technologies, major innovative issues and applicability to West African countries. *Renewable and Sustainable Energy Reviews*, 18, 306–315.

Salomoni, V.A., Majorana, C.E., Giannuzzi, G.M., Miliozzi, A., Di Maggio, R., Girardi, F., Mele, D., Lucentini, M. (2014). Thermal storage of sensible heat using concrete modules in solar power plants. *Solar Energy*, 103, 303–315.

Schmidt, M., Szczukowski, C., Roßkopf, L., Linder, M., Wörner, A. (2014). Experimental results of a 10 kW high temperature thermochemical storage reactor based on calcium hydroxide. *Appl. Therm. Eng.*, 62(2), 553–559.

Sequeira, C.A.C. (2019). *High Temperature Corrosion: Fundamentals and Engineering*. John Wiley & Sons, Hoboken.

Siegel, N.P. (2012). Thermal energy storage for solar power production. *Wiley Interdisciplinary Reviews: Energy and Environment*, 1(2), 119–131.

Steinmann, W.D. and Eck, M. (2006). Buffer storage for direct steam generation. *Solar Energy*, 80(10), 1277–1282.

Tescari, S., Singh, A., Agrafiotis, C., Oliveira, L., Breuer, S., Schlögl-Knothe, B., Roeb, M., Sattler, C. (2017). Experimental evaluation of a pilot-scale thermochemical storage system for a concentrated solar power plant. *Applied Energy*, 189, 66–75.

Thaker, S., Olufemi, Oni, A., Gemechu, E., Kumar, A. (2019). Evaluating energy and greenhouse gas emission footprints of thermal energy storage systems for concentrated solar power applications. *Journal of Energy Storage*, 26, 100992.

Warerkar, S., Schmitz, S., Goettsche, J., Hoffschmidt, B., Reißel, M., Tamme, R. (2011). Air-sand heat exchanger for high-temperature storage. *J. Sol. Energy Eng.*, 133(2), 021010.

Willmott, J.A. (2001). *Dynamics of Regenerative Heat Transfer*. Taylor & Francis, New York.

Winter, C.J., Sizmann, R.L., Vant-Hull, L.L. (1991). *Solar Power Plants*. Springer-Verlag, Berlin.

Xu, E., Wang, Z., Wei, G., Zhuang, J. (2012). Dynamic simulation of thermal energy storage system of Badaling 1 MW solar power tower plant. *Renewable Energy*, 39(1), 455–462.

Zaversky, F., Garcia-Barberena, J., Sanchez, M., Astrain, D. (2013). Transient molten salt two-tank thermal storage modeling for CSP performance simulations. *Solar Energy*, 93(7), 294–311.

Zhang, H., Benoit, H., Perez-Lopèz, I., Flamant, G., Tan, T., Baeyens, J. (2017). High-efficiency solar power towers using particle suspensions as heat carrier in the receiver and in the thermal energy storage. *Renewable Energy*, 111, 438–446.

List of Authors

Anthony DELAHAYE
INRAE
FRISE
Université Paris-Saclay
Antony
France

Stefania DOPPIU
CIC energiGUNE
BRTA
Vitoria-Gasteiz
Spain

Marie DUQUESNE
LaSIE, CNRS
La Rochelle Université
La Rochelle
France

Laurence FOURNAISON
INRAE
FRISE
Université Paris-Saclay
Antony
France

Erwin FRANQUET
Polytech'Lab
Université Côte d'Azur
Nice
France

Wahbi JOMAA
I2M, CNRS
Université de Bordeaux
Talence
France

Ana LAZARO
GITSE
I3A
University of Zaragoza
Spain

Philippe MARTY
LEGI, CNRS
Grenoble INP
Université Grenoble Alpes
Grenoble
France

Philippe MUGUERRA
Saipem
Montigny-le-Bretonneux
France

Lionel NADAU
ENGIE
Stains
France

Pierre ODRU
Previously IFPEN
Paris
France

Régis OLIVÈS
PROMES, CNRS
UPVD
Perpignan
France

Elena PALOMO DEL BARRIO
CIC energiGUNE
BRTA
Vitoria-Gasteiz
and
Ikerbasque
Bilbao
Spain

Index

E, F

G, H

I, K

L, M

N, O

P, R

S, T

U, V

W, X, Z

Summary of Volume 2

Chapter 3. Heat Storage Using Adsorption Processes

Larysa RATEL, Kevyn JOHANNES and Frédéric KUZNIK

Chapter 4. Heat Storage by Chemical Sorption Processes

Antoine PERRIGOT, Driss STITOU and Maxime PERIER-MUZET

Printed and bound by CPI Group (UK) Ltd, Croydon, CR0 4YY

13/09/2026

14960999-0004